BOTANICAL RESEARCH AND PRACTICES

PLANT ARCHAEOGENETICS

BOTANICAL RESEARCH AND PRACTICES

Additional books in this series can be found on Nova's website under the Series tab.

Additional E-books in this series can be found on Nova's website under the E-books tab.

BIOTECHNOLOGY IN AGRICULTURE, INDUSTRY AND MEDICINE

Additional books in this series can be found on Nova's website under the Series tab.

Additional E-books in this series can be found on Nova's website under the E-books tab

PLANT ARCHAEOGENETICS

GÁBOR GYULAI
EDITOR

Nova Science Publishers, Inc.
New York

LIBRARY OF CONGRESS CATALOGING-IN-PUBLICATION DATA

Plant archaeogenetics / editor: Gabor Gyulai.
    p. cm.
  Includes index.
  ISBN 978-1-61122-644-7 (hardcover)
  1. Paleobotany. 2. Plant morphology. 3. Plant genetics. I. Gyulai,
Gabor, 1953-
  QE905.P 2011
  561--dc22
                        2010041600

Published by Nova Science Publishers, Inc. † *New York*

CONTENTS

PREFACE

This book presents the most significant scientific information available in the study of plant achaeogenetics. A detailed examination is provided herein of plant achaeogenetics in order of the ages from ca. 30,000 years BP to the Middle ages. All fields of archaeogenetics are included from seed morphogenetics to aDNA extraction and analysis with molecular probes.

Chapter 1 – The identification and morphogenetics of ancient *Silene* seeds (*Caryophyllaceae*) is discussed in this chapter. Seeds were excavated from feeding chambers of ancient ground squirrels burrows (*Geomys*, subgenus *Urocitellus*) buried ca. 32 - 28,000 BP (*before present*) in the permafrost deposits of the Pleistocene Age (1.806 million years to 11,500 years BP), Kolyma lowland (Siberia). The ancient seeds were compared to seeds of extant species of *S. alba, S. chlorantha, S. nutans* and *S. stenophylla* presently growing in the same and neighboring regions. Using Scanning Electron Microscopy (SEM), the excavated *Silene* seeds were identified to be of *Silene stenophylla*.

Chapter 2 – Considerable quantities of plant remains of more than 40 taxa were characterized from the permanently frozen sediments of Northeast Siberia. Plant remains showed nearly uniform distribution with great numbers of seeds in the feeding chambers of ancient rodent burrows buried in permafrost. The degree of morphological integrity of seeds was good and identified to species. Distribution of species was indicative for the environmental and climate conditions of the Late Pleistocene (32 - 28,000 year ago) with vegetation of open larch forest (*Larix cajanderi*) and tundra.

Chapter 3 – Ohalo II is a submerged, late Upper Paleolithic (locally termed Early Epipalaeolithic) site, radiocarbon dated to ca 22,500–23,500 BP. (Kislev *et al.* 1992). The site is located on the southwestern shore of the Sea of Galilee (Lake Kinneret), Rift Valley, Israel (Figure 1). This hunter-gatherer-fisher camp covers more than 2000m^2 (0.2 ha) and includes the remains of six brush huts with numerous plant remains, open-air hearths, and a human grave. The site was occupied during the Last Glacial Maximum (LGM), a period of cold and dry climate when ice sheets covered parts of North America and Europe. Almost 100 taxa are represented among the identified plant remains (Nadel 1997; Simchoni 1998; Weiss 2009). The most numerous monocot and dicot species will be discussed in this Chapter.

Chapter 4 – Morphogenetics of plant remains excavated in the Carpathian Basin from 8000 BP (*before present*) to the late Medieval Age (17th cent. CE) are discussed in this chapter. The collection of samples goes back more than 130 years (1877 – 2010), consisting of 9 million seeds of 660 plant taxa from 460 archaeological sites representing the results of 24 researchers. As the Carpathian Basin is one of the longest continuously inhabited regions in Europe, the morphogenetical analysis of the most ancient seeds helps to trace archaeogenetical events, the origin of agriculture, weed history and environmental changes.

Chapter 5 – Morphogenetics of grape (*Vitis vinifera)* seed remains from Roman (3rd CENT CE) (*current ere*) and Medieval (11th to 15th CENT CE) archeological sites in Hungary are discussed in this Chapter. Seeds were analyzed by LM (Light Microscopy) and SEM (*Scanning Electron Microscopy*). Excavation sites included Budapest (*Aquincum*; 3rd CENT, Hungary) and Keszthely (Fenékpuszta; 5th CENT, Hungary) of Roman Age; and Győr (Ece, 11th-12th CENT, Hungary), Debrecen (13th CENT, Hungary) and the King's Palace of Árpád Dynasty at the Castle of Buda, Budapest (15th CENT, Hungary) of the Middle Ages. Ancient seeds were compared to thirty current grape varieties of similar seed size, shape and morphology. The current blue grape variety *Vitis vinifera* cv. '*kék bakator*' (*syn.:'Blue Bocca d'Oro*'; '*aranybogyó*', '*golden grapes*') (# 25, Figure 4) was found most similar in seed morphology to one of the ancient samples (15th CENT Debrecen, Hungary) which indicates the antiquity of this cultivar.

Chapter 6 – Ancient DNA (aDNA) was extracted from seed remains of common millet (*Panicum miliaceum*) excavated from sites of 4th CENT CE (Darhan, Mongolia) and 15th CENT CE (*current era*) (Budapest, Hungary). aDNA samples were analyzed by AFLP, SSR, ISSR, mtDNA and CAP-mtDNA methods.

Agarose gel electrophoresis revealed that extensive DNA degradation had occurred in the 4th CENT millet resulting in only 2 AFLP fragments (1.2 %), compared to the 15th CENT millet with 158 fragments (40.0 %) and current millet cv *Topáz* with 264 fragments (100 %). Eight AFLP fragments were sequenced after reamplification and cloned in the medieval millet.

Four microsatellites (SSR) were amplified in the medieval millet at the *gln*4, *sh*1, *rps*28 and *rps*15 gene loci and compared to current variety cv *Topáz*. No SSR was amplified in the 4th CENT sample. After sequencing, the medieval SSR fragments showed identical alignments with current millet sequences with only one SNP (*single nucleotide polymorphism*) at the 6th position (A to G) of the *rps*28 locus.

The mtDNA specific primer pair amplified an 1117 bp fragment at the 5S-18S rDNA locus in the 15th CENT millet and current millet cv '*Omskoje*'. In the CAP-analysis of this fragment, there were no observed changes in the DNA sequences at restriction sites of *Taq*I, *Bsu*RI, *Hinf*I, *Mbo*I, *Alu*I and *Rsa*I giving the same fragment pattern on agarose gel with the same numbers and lengths of digested fragments. After sequencing of *Mbo*I fragments of the 5S-18S rDNA, no SNP was observed in the nucleotide sequences. Authenticity of aDNA sequences were proved by BLAST analysis in each case.

In the ISSR analysis, seven ISSR primers and combinations amplified 15 alleles in the 15th CENT millet and twenty current millet cultivars used for comparative analysis. Phenotype reconstruction was carried out based on ISSR dendrogram, which revealed that medieval millet might had been very similar to current variety '*Omskoje*'.

Chapter 7 – Seed remains of watermelon (*Citrullus lanatus lanatus*) were excavated from two sites from the Middle Ages 13th (Debrecen) and 15th centuries (Budapest), Hungary, processed by floatation followed by seed sorting and identification. After seed morphological analysis aDNAs were extracted and analyzed at loci of twelve microsatellites (SSRs), two cpDNA, RAPD, ITS, and *lcyb* gene with a final aim of sequence recovery and phenotype reconstruction. For comparative analysis, an herbarium sample from the 19th CENT (Pannonhalma, Hungary) and forty-four current *Citrullus* cultivars were used. Molecular dendrogram based on microsatellite analysis revealed the closest similarity to current watermelon cultivars with red flesh (13th CENT) and yellow flesh (15th CENT) colors. The 170-yr-old herbarium sample showed close molecular similarity to citron melons (*Citrullus lanatus citroides*) which also reflects the importance of citron melon as fodder in Hungary.

Chapter 8 – ITS, SSR, RAPD and sequence analyses of *a*ncient melon DNA (aDNA) extracted from 600-yr-old seed remains recovered from the 15th CENT site in Budapest (Hungary) were analyzed. An aseptic incubation of seeds followed by ITS (*i*nternal *t*ranscribed *s*pacer) analysis was used to exclude the exogenously and endogenously contaminated seeds and to detect SNPs (*s*ingle *n*ucleotide *p*olymorphism) in ITS1-5.8S-ITS2 region (667 bp) of rDNA (*r*ibosomal DNA). SNPs of ITS1 were observed at the 94-95th bp (GC to RC, AC, AG, AS, GG, GS, RS); and of ITS2 at 414th bp (A-to-T substitution), 470th bp (T to Y or C), 610th bp (A to R or G) and 633rd bp (A-to-G transition). A medieval sample was compared to 47 current melon cultivars and landraces with a final aim of genotype and phenotype reconstruction of the medieval melon. For RAPD analysis, 60 primers from primer sets OP-A, -K and -M, (1 to 20 in each case) were applied resulting in a total of 105 fragments in the 15th CENT and current melons. Of them, 61 were monomorphic and 44 showed polymorphisms. Microsatellites at nuclear *s*imple *s*equence *r*epeats (SSR) were detected by *a*utomated *l*aser *f*luorometer (ALF). Eight of the twenty SSR primer pairs tested have amplified 40 microsatellite alleles with a total of 463 fragments in the medieval and current melons. The number of alleles per SSR loci ranged from 2 to 7 with an average of 5.7 including *Cm*CT44 (2 alleles), *Cm*AG59 (5 alleles), *Cm*GA104 (5 alleles), *Cm*CT134 (4 alleles), *Cm*TA134 (6 alleles), *Cm*CTT144 (7 alleles), *Cm*TC168 (6 alleles) and *Cm*CT170 (5 alleles). Sequence analysis of the SSR alleles at the dinucleotide (CT)n and trinucleotide (CTT)n loci showed different fragment lengths depending on changes in the number of core unit. The length of SSRs did not show a time dependent variation in lengths as 15th CENT melon showed SSRs of intermediate sizes compared to current cultivars. Molecular dendrogram, based on the presence versus absence of SSR alleles, revealed that medieval melon had the closest genetic similarity to a current melon cultivar '*Hógolyó*' (#24) of *inodorus* fruit type (also called winter melon, fist-size melon) with smooth yellow rind, and green flesh color. The results indicate the importance of winter melons in the medieval Europe. As the cultivation of melon in Europe started only in the 13th CENT, however, the Hungarians already named melon (*dinnye*) in the early 11th CENT, the melon samples recovered from the 15th CENT Budapest may originate from one of the oldest cultivated melons in Europe.

Chapter 9 – The Cucurbitaceae family is comprised of several economically important cultivated species including watermelon, melon, cucumber and squash. Members of the family have several unique traits such as lianous structure of the plant body, fleshy fruits and unusual mode of sex determination. Large scale sequencing analysis of many cucurbit species and almost all cucurbit genera indicated that the family probably originated in Asia and spread via transoceanic long distance dispersal to the different continents. While *Cucurbita* species are among the earliest plants domesticated in the New World, the bottle gourd might have been domesticated in Asia first and later in Africa. Watermelon was probably domesticated in Africa, the cucumber in India, while the origin of domestication of melon is still uncertain. Major efforts are underway to sequence several of the economically important cucurbit genomes and functional genomic studies. Together with archaeogenetical data, this will lead to a deeper understanding of the origin, and genetic basis of cucurbit crop domestication.

Chapter 10 – Delicate species, old varieties and heirlooms are threatened by extinction due to changes in environmental conditions, over cultivation, and falling below the minimum viable population. Conservation genetics provides effective tools of micropropagation to produce a large number of clones. As the clones develop from somatic tissues or organs, the genome (DNA content) remains identical in each clone (Gyulai *et al.* 2003). In the case of the 300-year-old *Robinia pseudoacaia cv. 'Bábolna'*, micropropagation was successfully used for clonal propagation resulting in identical clones. After rooting, clones will be planted in an archaeogenetical garden of botany in Gödöllő (Hungary).

Chapter 11 – The aim of this Chapter is to compare thirteen different DNA extraction methods for the isolation of PCR amplifiable aDNAs from mummified soft tissues recovered from 1731-1841 (Vác, Hungary). Among the techniques and methods, the silicabased *guanidium thiocyanate* treatment and the phenol/chloroform extraction method combined with PTB (*N-phenacyl-thiazolium bromide*) were found to be powerful for aDNA isolation. These results might be indicative for Plant Archaeogenetists.

Chapter 12 – The computer-assisted morphometry of two-dimensional and three-dimensional size, shape, and other morphological parameters are used in the analysis of seed morphology and variation to discriminate wild and domesticated population. An index RGiD (the *R*ovner-*G*yulai *I*ndex of seed *D*omestication) was developed (Rovner and Gyulai 2007) based on the comparative analysis of wild and domesticated forms of wheats discussed in this Chapter.

Chapter 13 – Artistic works from prehistory and antiquity to the present constitute an alternate source of information on crop plants that become a valuable resource for investigations involving genetic and taxonomic information, as well as crop history. Sources of plant iconography include cave paintings, ancient mosaics, sculpture, carvings and inlays, frescos, tapestries, illustrated manuscripts, herbals, and books, as well as photographs of desiccated remnants.

I. Morphogenetics

In: Plant Archaeogenetics
Editor: Gábor Gyulai, pp. 3-9

ISBN: 978-1-61122-644-7
© 2011 Nova Science Publishers, Inc.

Chapter 1

MORPHOGENETICS OF *Silene stenophylla* SEEDS FROM PERMAFROST OF THE LATE PLEISTOCENE (32 - 28,000 BP)

Gábor Gyulai[1,], Lilja Murenyetz[1,2,3],*
Zsigmond G. Gyulai[1,4], Viacheslav L. Stakhov[1,3],
Svetlana G. Yashina[2] and Stanislav V. Gubin[2]

[1]Institute of Genetics and Biotechnology, St. Stephanus University,
Gödöllő 2103, Hungary
[2]Institute of Physicochemical and Biological Problems in Soil Science,
Russian Academy of Sciences, Pushchino 142290, Russia
[3]Institute of Gene Biology, Russian Academy of Sciences, Moscow 119334, Russia
[4]Török Ignác High School, Sándor Petőfi 12, Gödöllő 2100, Hungary

ABSTRACT

The identification and morphogenetics of ancient *Silene* seeds (*Caryophyllaceae*) is discussed in this chapter. Seeds were excavated from feeding chambers of ancient ground squirrels burrows (*Geomys*, subgenus *Urocitellus*) buried ca. 32 - 28,000 BP (*before present*) in the permafrost deposits of the Pleistocene Age (1.806 million years to 11,500 years BP), Kolyma lowland (Siberia). The ancient seeds were compared to seeds of extant species of *S. alba, S. chlorantha, S. nutans* and *S. stenophylla* presently growing in the same and neighboring regions. Using Scanning Electron Microscopy (SEM), the excavated *Silene* seeds were identified to be of *Silene stenophylla*.

* E-mail: gyulai.gabor@mkk.szie.hu

INTRODUCTION

The ancient burrows of ground squirrels (*Geomys*, subgenus *Urocitellus*), buried in permafrost deposits of the Late Pleistocene age at the site of Kolyma (Siberia), provided unique seed materials for evolution analysis (Gubin and Khasanov 1996; Gubin *et al.* this volume). These burrows with seed materials in their feeding chambers have been dated back to 32 - 28,000 years BP determined by radiocarbon (Yashina *et al.* 2002; Stakhov *et al.* 2008). It is supposed that burrows have not thawed out from freezing temperature by now.

The Pleistocene has been dated from 1.806 million years (+/- 5.000) to 11,500 years BP, expressed in radiocarbon years. The Pleistocene climate was characterized by repeated glacial cycles with a maximum glacial extent when 30% of the Earth's surface (namely permafrost) was covered by ice (today, approximately 20% of the Earth's is covered by permafrost) (Kislev *et al.* this volume). The mean annual temperature at the edge of the ice was -6 °C, and at the edge of the permafrost, 0°C. Research evidence indicates that humans evolved into their present form during the Pleistocene along with the major extinction events of the Neanderthals and large animals such as mammoths, mastodons, saber-toothed cats, etc. The extinctions were especially severe in North America where native horses and camels became extinct.

Archaeological samples preserved under optimal conditions at low (or permafrost) temperature (Suh *et al.* 2000; Willerslev *et al.* 2003; Schlumbaum *et al.* 2008) can supply aDNA with amplifiable quality as shown in the studies of 15-20 thousand year-old cereals (rice, wild wheat, barley) (Suh *et al.* 2000; Özkan *et al.* 2002) and medieval plants (Gyulai *et al.* 2001, 2006; Bisztray *et al.* 2004; Lágler *et al.* 2005; Szabó *et al.* 2005), or, in the case of fossilized samples, the deoxyribose backbone of aDNA as shown in 55 million year-old (Lower Eocene) *Myrtaceae* fossils (Ozerov *et al.* 2006). The aDNA analyses of ancient *Silene* seeds of present study are in progress.

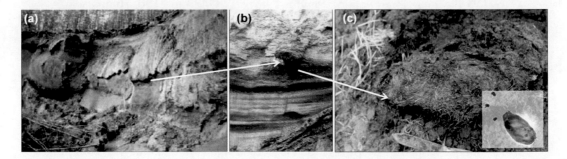

Figure 1. Excavations *(a, b)* of gopher (*Geomys* ssp.) holes *(c)* buried under a Pleistocene-age permafrost (32 - 28,000 BP) at a site near Kolyma, Siberia. Ancient *Silene* ball and seeds are enlarged. For dimensions notice the trees on the top *(a)* and a knife *(c)*.

MATERIALS AND METHODS

The *Silene* seeds in this study were excavated in the Kolyma region (Siberia) at the famous mammoths excavation site (Stakhov *et al.* 2008) (Figure 1. (1) #P1075 and (2) #P1300). Radio carbon analysis was carried out according to the basic methodology of

Arnold and Libby (1949) by Yashina *et al.* (2002) and Stakhov *et al.* (2008). Sediment samples were processed by seed sorting and identification in the laboratory according to Schermann (1966) and Gyulai *et al.* (2006). For SEM (*Scanning Electron Microscopy*) analysis, seeds were air dried, fixed in glutaraldehyde (5% w/v in phosphate buffer 0.07 M, pH 7.2) and washed three times in the same buffer for 10 minutes. Samples were desiccated in an acetone concentration series (10-50-70-90-100%), dehydrated at the CO_2 critical point (Blazers CDC 020), and covered with gold (30 nm). Samples were examined and photographed using a TESLA BS-300 scanning electron microscope. For LM (*Light Microscopy*) analysis, a Leica microscope (# 301-371.010) was used (Basli *et al.* 2009). For comparative analyses, botanical seed samples of extant *Silene* species of *S. alba, S. chlorantha, S. nutans* and *S. stenophylla* were treated the same way.

CONCLUSION

The study of ancient plant fossils and remains by *arhaeo/paleo botany*, and the study of aDNA (ancient DNA) by *archaeo/paleo genetics* supplies unique data to evaluate changes in genetic variation and domestication (Özkan *et al.* 2002) that occurred during evolution over the past hundreds to million years (Gugerli *et al.* 2005; Gyulai *et al.* 2006).

Fossilized samples of *Bangiomorpha pubescens* (a red alga) from Canada prove that chloroplasts originated more than 1.2 billion years ago (Butterfield 2000). Fossilization, coupled by charcoalification, left floral morphology of ancient plants perfectly preserved at a site in Europe (ca. 120 million years BP) (Friis *et al.* 2001), in Sayreville (NJ, USA) from the earliest Upper Cretaceous time (ca. 90 million years BP) (Gandolfo *et al.* 2004; Crepet *et al.* 2004), and in the Dakota Formation of Kansas (Kovach and Dilcher 1988). Fossils of basal angiosperms (*Archaefructus sp*) were also discovered from the lower early Cretaceous period in China (Zhou *et al.* 2003). Extinct and petrified angiosperm species (e.g. *Pinus tuzsoni* Greguss; *syn. Pinuxylon tarnocziense* Tuzson) were identified from a 20 million year-old (Lower Miocene) site at Ipolytarnóc (Hungary) (Andreánszky 1996; Greguss 1972; Erdei *et al.* 2007; Hably 2006; Süss 2007).

Radiocarbon age dating is generally used to determine the age of carbonaceous materials up to about 50 - 60,000 years based on the naturally occurring isotope carbon-14 (^{14}C) (Plastino *et al.* 2001). The technique was developed by Libby (Arnold and Libby 1949), who was awarded the Nobel Prize in 1960. The methodology of radiocarbon dating is based on the fact that carbon has two stable, nonradioactive isotopes (^{12}C and ^{13}C); and one unstable isotope (^{14}C) with a half-life of 5,568 ± 30 years (expressed in Libby half-life) or 5730 years (in Cambridge half-life). Practically, the small amount of ^{14}C would have vanished from the Earth a long time ago, except for the cosmic rays which enter the atmosphere and continuously generate it from nitrogen molecules (N_2) according to the classical isotope reaction, as n (neutron) + $^{14}N_7 \rightarrow ^{14}C_6$ + p (proton). The highest rate of ^{14}C production takes place at altitudes of 9 to 15 km but it spreads evenly throughout the atmosphere, producing at a constant rate and with the proportion of radioactive to non-radioactive carbon also remaining constant, ca. 1 ^{14}C / 600 billion atoms/mole. As nonradioactive C-isotopes ^{14}C also react with oxygen to form CO_2, which enters plants by photosynthesis, and from plants, it is

Figure 2. SEM morphology of ancient seeds of *S. stenophylla* (*1 – 2*) (32 – 28,000 BP), and compared to current *Silene* seeds at three magnifications ranges (a – b). (*3*) *S. stenophylla* (Kolyma). (*4*) *S. alba* (Moscow). (*5*) *S. chlorantha* (Voronyezs). (*6*) *S. nutans* (Moscow). (*7*) *S. viscosa* (Moscow).

incorporated into animal tissue. When organisms (plants or animals) die, the incorporation of ^{14}C stops and its content gradually decreases in the cadaver through radioactive decay by turning back the generative reaction producing $^{14}N_7$ according to the beta-minus decay reaction: n (neutron) + $^{14}C_6 \rightarrow {}^{14}N_7 + e^-$ (electron) + v_e (anti neutrino). This decay is used to measure how long ago a piece of once-living material died and this is expressed as years (calibrated as 1950 CE) (*Current Era*).

The approximate age of the ancient *Silene* (*Caryophyllaceae*) seeds of the present study were determined by radiocarbon method to be 32 - 28,000 years old.

Ancient *Silene* seeds were compared to seed samples of four recent species growing in the same region (*S. alba*, *S. chlorantha*, *S. nutans* and *S. stenophylla*) and determined to be of *Silene stenophylla* (Ledeb.) by SEM and LM (Figure 2). The ancient *Silene* seeds had morphological features characteristic of those of contemporary *S. stenophylla* seeds, except for being of smaller size (Figure 2). Interestingly, the ancient seeds had damaged embryos (Figure 2c), which might be the result of the gophers' activity and effort to prevent undesired germination in the burrows.

To conclude, the ancient burrows of ground squirrels (*Geomys*, subgenus *Urocitellus*) buried in permafrost deposits of Late Pleistocene age provides unique seed materials for archaeogenetical studies (Zazula *et al*. 2003; Gubin *et al*. this volume). As cells of well-preserved permafrost seeds might carry intact cells with intact aDNA, an experiment for callus initiation for plant regeneration in tissue culture is in progress; similar to the former successful (Aufhammer and Fischbeck, 1964; Ruckenbauer 1971), unsuccessful (Szabó *et al*. 2005; Lágler *et al*. 2005) and doubtful ancient seed germination results (Porshild *et al*. 1967; Quinn, 1999; Shen-Miller 2002).

ACKNOWLEDGMENT

Ancient seeds were kindly provided by Dr. SV Maksimovich (Pushchino, Russia).

REFERENCES

Andreánszky G (1966) The Upper Oligocene Flora of Hungary. Analysis of the site at the Wind Brickyard, Eger. *Studia Biologica Hungarica* 5: 1–151.

Aufhammer G and Fischbeck G (1964) Ergebnisse von Gefass- und Feldversuchen mit dem Nachbau keimfahiger Gersten- und Haferkorner aus dem Grundstein des 1832 errichteten. Nürnberger Stadttheates. *Zeitschrift für Pflanzenzuchtung* 51: 345-378.

Arnold JR and Libby WF (1949) Age Determinations by Radiocarbon Content: Checks with Samples of Known Age. *Science* 110: 678–680.

Başli AG, Gyulai G, Tóth Z, Güner A, Szabó Z, Stakhov VL, Murenyetz L, Yashina SG, Heszky L, Gubin SV (2009) Light and Scanning Electron Microscopic Analysis of *Silene stenophylla* Seeds Excavated from Pleistocene-Age (Kolyma). *Anadolu Univ J Sci and Technol* 10: 161-167.

Bisztray GyD, R Bacsó, P Bodor, G Facsar, F Gyulai, I Velich (2004) Archaeobotanical and genetical methods to analyse 600-years-old seeds of horticultural plants. Book of

Abstracts 5th IVCHB Symposium, In Vitro Culture and HorticulturaBreeding, 12-17. September 2004, Debrecen, Hungary, p. 212.

Butterfield NJ (2000) *Bangiomorpha pubescens* n. gen., n. sp.: implications for the evolution of sex, multicellularity and the Mesoproterozoic/Neoproterozoic radiation of eukaryotes. *Paleobiology* 26: 386-404.

Crepet W, Nixon KC, Gandolfo MA (2004) Fossil evidence and phylogeny: the age of major angiosperm clades based on mesofossil and macrofossil evidence from cretaceous deposits. *Amercan Journal of Botany* 91: 1666–1682.

Erdei B, Hably L, Kázmér M, Utescher T, Bruch AA (2007) Neogene flora and vegetation development in the Pannonian Basin - relations to palaeoclimate and palaeogeography. – *Palaeogeography, Palaeoclimatology, Palaeoecology* 253: 131-156.

Friis EM, Pedersen KR, Crane PR (2001) Fossil evidence of water lilies (Nymphaeales) in the Early Cretaceous. *Nature* 410: 357-360.

Gandolfo MA, Nixon KC, Crepet WL (2004) Cretaceous flowers of *Nymphaeaceae* and implications for complex insect entrapment pollination mechanisms in early Angiosperms. *Proceedings of the National Academy of Sciences USA* 101: 8056-8060.

Greguss P (1972) Xylotomy of the living conifers. Pp. 329, Akadémiai Kiadó, Budapest.

Gubin SV and Khasanov BF (1996) Fossil Burrows of Mammals in the Loess-Ice Deposits of the Kolyma-Indigirka Lowland, Doklady AN, 1996, vol. 346, no. 2, pp. 278–279 [Doklady Biol. Sci. (Engl. Transl.), vol. 346, pp. 26–27].

Gugerli F, Parducci L, Petit RJ (2005) Ancient plant DNA: review and prospects. New *Phytologist* 166: 409-418.

Gyulai G, A Magda, J Kiss, F Gyulai, L Holly, L Heszky (2001) DNS izolálás és PCR-amplifikáció 700 éves növény magvakból. VII. Növénynemesítési Tudományos Napok, Budapest, p.89.

Gyulai G, Humphreys M, Lágler R, Szabó Z, Tóth Z, Bittsánszky A, Gyulai F, Heszky L (2006) Seed remains of common millet from the 4th (Mongolia) and 15th (Hungary) centuries; AFLP, SSR, and mtDNA sequence recoveries. *Seed Science Research* 16: 179-191.

Hably L (2006) Catalogue of the hungarian cenozoic leaf, fruit and seed floras from 1856 to 2005. *Studia Botanica Hungarica* 37: 41–129.

Kovach WL and DL Dilcher (1988) Megaspores and Other Dispersed Plant Remains from the Dakota Formation (Cenomanian) of Kansas, U.S.A. *Palynology* 12: 89-119.

Lágler R, Gyulai G, Humphreys M, Szabó Z, Horváth L, Bittsánszky A, Kiss J, Holly L, Heszky L (2005) Morphological and molecular analysis of common millet (*P. miliaceum*) cultivars compared to an aDNA sample from the 15th century (Hungary). *Euphytica* 146: 77-85.

Ozerov IA, Zhinkina NA, Efimov AM, Machs EM, Rodionov AV (2006) Feulgen-positive staining of the cell nuclei in fossilized leaf and fruit tissues of the Lower Eocene Myrtaceae. *Botanical Journal of Linnean Society* 150: 315-321.

Özkan H, Brandolini A, Schäfer-Pregl R, Salamini F (2002) AFLP analysis of a collection of tetraploid wheats indicates the origin of emmer and hard wheat domestication in southeast Turkey. *Molecular Biology and Evolution* 19: 1797–1801.

Porsild AE, Pharington CR, Mulligan GA (1967) '*Lupinus arcticus*' Wats. grown from seeds of Pleistocene age. *Science* 158: 113–114.

Plastino W, Kaihola L, Bartolomei P, Bella F (2001) Cosmic background reduction in the radiocarbon measurement by scintillation spectrometry at the underground laboratory of Gran Sasso. *Radiocarbon* 43: 157–161.

Quinn RM (1999) Kamut: Ancient grain, new cereal. In: *Perspectives on new crops and new uses.* Ed: J. Janick, pp. 182–183, ASHS Press, Alexandria, VA.

Ruckenbauer P von (1971) *Keimfähiger Winterweizen aus dem Jahre 1877.* Beobachtungen und Versuche. pp. 372-386. Inst. f. Pflanzenbau und Pflanzenzuchtung d. Hochschule f. Bodenkultur in Wien.

Schermann Sz (1966) Magismeret (*Seed morphology*), Vol. I (pp. 861) and II (pp. 209) Akadémiai Kiadó, Budapest.

Schlumbaum A, Tensen M, Jaenicke-Despres V (2008) Ancient plant DNA in archaeobotany. *Vegetation History and Archaeobotany* 17: 233–244.

Shen-Miller J (2002) Sacred lotus, the long-living fruits of China Antique. *Seed Science Research* 12: 131–143.

Suh HS, Cho JH, Lee YJ, Heu MH (2000) RAPD variation of 13,010 and 17,310 year-old carbonized rice. *4th International Rice Genetics Symposium*, Manilla, Philipines, Oct. 22-27.

Stakhov VL, Gubin SV, Maksimovich SV, Rebrikov DV, Savilova AM, Kochkina GA, Ozerskaya SM, Ivanushkina NE, Vorob'eva EA (2008) Microbial Communities of Ancient Seeds Derived from Permanently Frozen Pleistocene Deposits. *Microbiology* 77: 348–355.

Süss H (2007) Wood fossils of the morphogenus *Spiroplatanoxylon* gen. nov. from the Tertiary of Europe and the Middle East. *Feddes Repertorium* 118:1-19.

Szabó Z, Gyulai G, Humphreys M, Horváth L, Bittsánszky A, Lágler R, Heszky L (2005) Genetic variation of melon (*C. melo*) compared to an extinct landrace from the Middle Ages (Hungary) I. rDNA, SSR and SNP analysis of 47 cultivars. *Euphytica* 146: 87-94.

Zazula GD, Froese DG, Schweger CE, Mathewes RW, Beaudoin AB, Telka AM, Harington CR, Westgate JA (2003) Ice age steppe vegetation in East Beringia. *Nature* 423: 603.

Zhou Z, Barrett PM, Hilton J (2003) An exceptionally preserved Lower Cretaceous ecosystem. *Nature* 421: 807-814.

Willerslev E, Hansen AJ, Binladen J, Brand TB, Gilbert MTP, Shapiro B, Bunce M, Wiuf C, Gilichinsky DA, Cooper A (2003) Diverse plant and animal genetic records from Holocene and Pleistocene sediments. *Science* 300: 791–795.

Yashina SG, Gubin SV, Shabaeva EV, Egorova EF, Maksimovich SV (2002) Viability of Higher Plant Seeds of Late Pleistocene Age from Permafrost Deposits as Determined by *in vitro* Culturing, Doklady AN, 2002, Vol. 383, Nº 5, 714–717 (Doklady Biol. Sci. (Engl. Transl.), Vol. 383, 167–170).

In: Plant Archaeogenetics

Editor: Gábor Gyulai, pp. 11-21

ISBN: 978-1-61122-644-7

© 2011 Nova Science Publishers, Inc.

Chapter 2

MORPHOGENETICS OF PLANT REMAINS FROM PALEOSOLS AND RODENT BURROWS BURIED IN PERMAFROST OF THE LATE PLEISTOCENE (32 - 28,000 BP)

Stanislav V. Gubin[1,], Stanislav V. Maximovich[1],*
Oksana G. Zanina[1] and Viacheslav L. Stakhov[1,2]
[1]Institute of Physicochemical and Biological Problems in Soil Science,
Russian Academy of Sciences, Pushchino 142290, Russia
[2]Institute of Gene Biology, Russian Academy of Sciences, Moscow 119334, Russia

ABSTRACT

Considerable quantities of plant remains of more than 40 taxa were characterized from the permanently frozen sediments of Northeast Siberia. Plant remains showed nearly uniform distribution with great numbers of seeds in the feeding chambers of ancient rodent burrows buried in permafrost. The degree of morphological integrity of seeds was good and identified to species. Distribution of species was indicative for the environmental and climate conditions of the Late Pleistocene (32 - 28,000 year ago) with vegetation of open larch forest (*Larix cajanderi*) and tundra.

PERMAFROST SEDIMENTS

Permanently frozen (termed *permafrost*) sediments of ice complexes were widely spread in the northeastern part of Eurasia, Alaska and Canada. These sediments, known as Yedoma sediments, or Yedoma, had formed during the late Pleistocene age (between 70 to 10,000 BP) (*before present*). Yedoma is represented by many meters of frozen pulverescent and/or sandy-

---

[*] E-mail: gubin@issp.serpukhov.su

pulverescent sediments with pearl-gray and/or brownish colors and usually split by prolonged ice veins, sometimes up to 6-8 meters thick. The mass of sediment contains local patches of buried soils, alluvial deposits and peat. Yedoma sediments are similar to European loess in structure and composition [1, 10]. Hypothetically, loess formed on vast plain areas under conditions of severe and nearly snowless winters and relatively warm but short summers, with superficial occurrence of permafrost at the depth about 0.6 – 0.8 meters under the surface, and moderate dust precipitations on the contemporary surface [3, 16]. With time, mineral matter accumulated on the contemporary surface, causing the level of permafrost to rise as well. Incoming minerals and organic residues of plants and animals were partly modified by processes of soil formation before transition to a buried stage and subsequently to permafrost. Syngenetic soil formation process (the terms refers to the combination of continuous mineral matter deposition on the contemporary surface and its partial modification by soil-forming processes) did not generate apparent soil horizons. The permanently frozen matter formed by the combination of syngenetic soil-formation and freezing is referred to as cryopedolith as of Yedoma.

After the active soil layer passed into a permanently frozen state, all of the organic compounds interned, including plant bodies and seeds, have been stored for thousands of years under a slightly fluctuating temperature range from -10 to -20 °C. Such a low temperature favored preservation as one of the main reasons Yedoma sediments still contain such amounts of well-distinguishable organic compounds; sometimes nearly in their original morphological state. Different well-preserved organisms have been found in cryopedolith strata, such as microbes, plant seeds, moss spores, aerial parts of herbs (mainly of cereals and sedges), shrubs and semifrutex, earthnut parts of plants *in situ,* and also bones of mammoth-age fauna such as mammoth, woolly rhinoceros, bison, musk buffalo, horse, and rodents, etc. [3, 5, 6, 9, 12, 13, 14].

Occasionally, during the Pleistocene, the overall arid and cold climatic conditions were interlaced with humid periods for centuries, which generated interstratified layers of graded soils in the bulk of cryopedolith. These soils typically contain clay and humic layers which constitute evidence of swamp formation processes.

PLANT REMAINS IN ANCIENT RODENT BURROWS

We have examined the content of more than 20 burrows of ground-squirrels (*Spermophilus parryii*). Burrows located on mountainsides contained greater amounts of seeds than those in lowland [4]. Some burrows were occasionally discovered in the outcroppings of permafrost within layers of buried *paleosols*, and some within the freshly formed upright walls of thermokarst steeps in the depth of 15-30 meters below the today surface. Such burrows were found not only in the riverbanks and lakesides of North Yakutia and Kolyma-lowland, but in Alaska (USA), and Yukon territories of Canada [4, 16, 17, 24, 25].

For sampling, summer time (July-August) is the best period as active thawing and erosion brings new portions of permafrost to the surface. In some cases, the whole space of the feeding chamber was filled with well-milled plant remains and wool. More often, the feeding chambers also contained fragments of plant stems and leaves, spikelets of cereals, springs of semifrutex and shrubs, remains of insects, puparium of flies, pieces of bird eggshell, excrement, bones of various rodents (mainly of ground squirrels and mice) and even their semi-decomposed corpses, etc. The feeding chambers of ancient burrows often contained substantial amounts of plant seeds and fruits. Sometimes the average content was as high as 600 - 800 thousand fragments per feeding chamber [4]. Low temperatures (about -10°C) of enclosing sediments retarded decay of ancient seeds, readily allowing taxonomic identifications. The age of these seeds can be confirmed by carbon-14 dating, as the specimens of the chapter presented were dated to 32-28,000 years old [4, 7, 8, 22].

As permafrost acts as a natural cryo-depositary for the perishable organic materials, deposits were found to be saturated with semidormant life. A wide variety of viable organisms of bacteria, algae, fungi, and protozoa were found in old permafrost deposits at the depth of dozens of meters below the surface [11, 18, 19, 26]. Moss spores and plant cells with certain metabolic activities were also reported [20, 21].

Burrows of different rodents were different in their structure and composition of feeding materials, and also in the condition of the contained materials. The degree of preservation of fossil seeds in ancient burrows depended on the general disposition of the burrows within permafrost strata, the structure and composition of the bearing strata, the general disposition of the seeds in the feeding chamber, initial frost penetration velocity, and the initial maturity of the seeds at the time of collection.

Extant species, *Spermophilus parryii,* reside in the well-drained territories along the sea-, river- and lake-sides of Yakutia. Structural analysis of the recent burrows showed that the feeding chambers are typically situated right above the permafrost level, where the average temperature fluctuates slightly above 0°C, even in summer. Nowadays, probably similar to conditions several tens of thousands of years ago, coolness and dryness of the atmosphere in these feeding chambers highly facilitated the forage preservation. Since the permafrost level gradually increased, the feeding chambers were soon interned in permafrost deeper and deeper and the feeding chambers became stored at a constant temperature range from -10 to -20°C without thawing from that moment to the present.

SEED SAMPLES FROM ANCIENT BURROWS

Fossilized seeds were typically located in the central part of feeding chambers and form a voidless mass cemented by ice. Colors of most seeds were unnaturally too dark, which indicates some decaying processes (Figure 1, 2). Seeds were easily identified by morphogenetic characters [2, 8].

Morphological parameters of seed shape, venation and pubescence pattern, and distribution of pores were used for identification. Seeds of *Poa* species were typically characterized by a light-colored coma at the basement of the glume. Seeds of other cereals were indicative by aristae on the surface. Heads of *Hordeum jubatum* were characteristic with simple and bifurcated awns with seeds between them. This morphology was discriminative from *Hordeum brevisubulatum*. Seeds of *Gentianopsis barbata* were usually found enclosed in characteristic fruitcases that facilitates the identification. Fruitcases of *Silene stenophylla* were also discriminative with a better degree of preservation and more natural colors. *Carex* seeds were enclosed in sacculi which were significant for the identification at genus level.

Seeds of *Hedysarum arcticum* were also easily distinguishable as enclosed in characteristic fruitcases (Table 1).

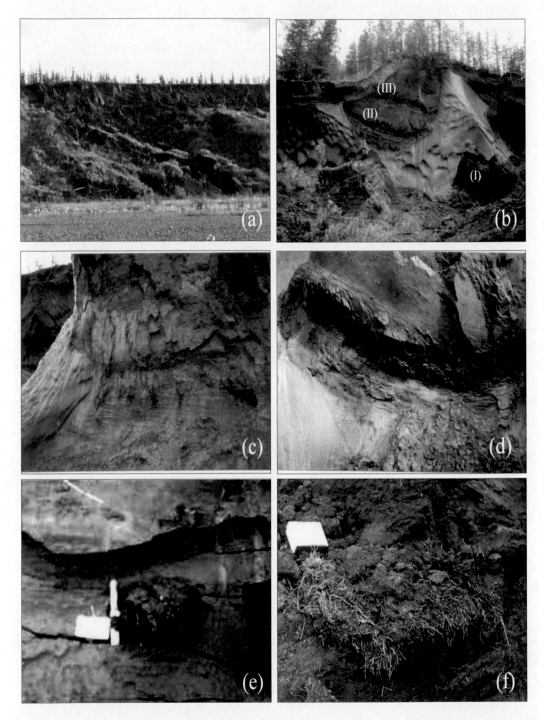

Figure 1. General view of the permafrost at Kolyma (Siberia) *(a)*. Paleosols of different ages *(I)* 42,200 ± 800 yr old (GIN 12872); *(II)* 36,500 ± 700 yr old (GIN 12873) *(b)*. Profiles of paleosols *(c, d)*. Fossilized burrows (32-28,000 yr old) with plant remains *(e, f)*.

Figure 2. Selected plant fossils from 32-28,000 yr old rodent burrows. *(a)* Plant and seed remains in permafrost paleosols (Kolyma, Siberia). *(b)* Spike of *Hordeum brevisubulatum*. *(c)* Balls of *Bistorta vivipara*. *(d, e)* Balls and seeds of *Silene stenophylla*. *(f)* Seeds of *Hedysarum arcticum*. *(g)* Parts of *Carex* sp. *(h, i)* Seed of *Kobresia myosuroides*. *(j)* Spikelets of *Poa attenuata* and *Poa botryoides*.

Table 1. Plant remains (+) by species from feeding chambers of ancient rodent burrows buried in permafrost (A – Open Woodland; B – Tundra)

Species	(A)	(B)	Species	(A)	(B)
Pioneer sp. of wet ecotope			Aulacomnium turgidum	+	+
Equisetum fluviatile	+		A. Palustre var. imbricatum	+	+
Arctophila fulva	+	+	Bryum arcticum	+	+
Rumex aquaticus	+	+	Cinclidium latifolium	+	+
Kobresia myosuroides	+	+	Hylocomium splendens	+	+
R. sibiricus	+	+	Meesia triquetra	+	+
Ranunculus repens	+	+	Polytrichum hyperboreum	+	+
Pioneer sp. of dry ecotope			P. jensenii	+	+
Arnica intermedia	+	+	*Lichens*		
Artemisia gmelinii	+		Cetraria cucullata	+	+
A. tilesii	+	+	C. laevigata	+	+
Tanacetum vulgare	+		Peltigera aphtosa	+	+
T. lateritum	+	+	*Wetland forest - tundra*		
T. macilentum	+	+	Betula exilis	+	+
Tripleurospermum hookeri	+	+	Carex appendiculata	+	+
Draba cinerea	+	+	C. atherodes	+	
D. pilosa	+	+	C. bonanzensis	+	
Sisymbrium polymorphum	+		C. concolor	+	+
Cerastium maximum	+	+	C. juncella	+	
Chenopodium album	+	+	C. lugens	+	+
Ch. prostratum	+	+	C. misandra	+	+
Corispermum elongatum	+		C. pallida	+	
Equisetum arvense	+	+	C. saxatilis	+	+
E. pratense	+		C. vesicata	+	
Chamaenerion angustifolium	+	+	Calamagrostis purpurea	+	+
Ch. latifolium	+	+	Elymus jacutensis	+	+
Plantago canescens	+		Hordeum brevisubulatum	+	+
Hordeum jubatum	+		Puccinella hauptiana	+	+
Potentilla stipularis	+	+	Polemonium acutiflorum	+	+
Plants of open larch forest			Bistorta vivipara	+	+
Vascular plants			Rubus hamaemorus	+	+
Cnidium cnidiifolium	+		Sanguisorba officinalis	+	+
Betula divaricata	+		Salix krylovii	+	+
Duschekia fruticosa	+		Pedicularis gymnostachya	+	+
Myosotis asiatica	+	+	P. kolymensis	+	
Rhodiola rosea	+	+	*Steppe plants*		
Empetrum androgynum	+	+	Allium strictum	+	+
Arctous alpina	+	+	Eritrichium sericeum	+	
Vaccinium uliginosum	+	+	Dianthus versicolor	+	
Vaccinium vitis-idaea	+		Eremogone tschuktschorum	+	+
Astragalus alpinus	+	+	Lychnis sibirica	+	+
Hedysarum arcticum	+	+	Silene repens	+	+
Gentianopsis barbata	+		S. stenophylla	+	+
Luzula multiflora	+	+	Dracocephalum palmatum	+	+
Larix cajanderi	+		Papaver nudicaule	+	
Pinus pumila	+		Calamagrostis purpurascens	+	+
Alopecurus alpinus	+	+	Elytrigia villosa	+	
Arctagrostis latifolia	+	+	Festuca lenensis	+	+
Bromopsis pumpelliana	+	-	Poa attenuata	+	+
Festuca altaica	+	+	P. botryoides	+	+
Poa alpigena	+	+	Androsace septentrionalis	+	+
P. sibirica	+	+	*Mountains and plain tundra*		
Potentilla nivea	+	+	Vaccinium minus		+
Salix glauca	+	+	Papaver polare		+
S. pulchra	+	+	Puccinella borealis		+
Kobresia myosuroides	+	+	Dryas octopetala ssp. subincisa		+
Mosses			Salix polaris		+
Aulacomnium palustre	+	+			

Fruitcases break down into spherical segments containing only one seed each at the stage of maturity. *Potentilla* seeds were also characteristic (Table 1). *Bistorta vivipara* was identified by the form of a stem instead of seeds with pear-shaped appearances and the peculiar dotted pattern on the surface.

Table 2. Plant remains from paleosols (pieces and *g** of remains of soil sample unit)

Species	Sampling layers										
	1310 ATC	1310 AT	1310 ATB	1310 CG	1321	1332 ATC	1332 AT	1333 A1	1333 AC	AT AT	AT AT
Moss Aulacomnium palustre	-	-	-	-	-	-	-	-	-	-	3
Moss Aulacomnium imbricatum	-	-	-	-	-	-	-	-	-	-	1
Arctagrostis latifolia (culm) *	-	-	-	0.03	-	-	-	-	-	-	-
Arctous alpine (seeds)	-	-	-	4	-	-	4	9	25	1	-
Aconogonon tripterocarpum (seeds)	-	-	-	-	-	-	6	-	-	--	
Betula divaricata (wood)*	-	-	-	-	-	-	-	0.1	-	-	-
Bistorta elliptica (seeds)	1	1	1	-	-	-	3	11	-	1	-
Carex appendiculata (seeds)	-	8	43	40	38	-	4	77	-	301	-
Carex aterrima (seeds)	-	-	-	-	-	-	-	-	-	3	-
Carex concolor (seeds)	8	13	95	116	12	3	6	120	-	60	-
Carex juncella (seeds)	-	-	10	15	2	-	1	4	-	-	-
Carex rhynchophysa (seeds)	-	-	-	-	-	-	-	-	-	-	1.0
Carex sp. (roots)	-	-	-	-	-	-	-	-	-	4.0	-
Comarum palustre (stem particles) * 1.3	-	-	-	-	-	-	-	-	-	-	-
Comarum palustre (seeds)	-	-	-	-	-	-	-	-	-	-	30
Larix cajanderi (wood / leaves) * 0.1	-	-	-	-	-	-	-	-	-	-	0.2.
Larix cajanderi (seeds)	-	-	-	-	-	-	-	-	-	-	6
Potentilla stipularis (seeds)	5	2	10	29	-	3	70	2	-	-	-
Puccinella sp. (seeds)	-	-	-	-	-	-	-	-	-	-	-
Kobresia myosuroides, (seeds)	-	-	-	-	1	-	-	-	-	-	-
Salix ssp. (wood) *	-	-	-	-	0.2	3.2	32.9	3.5	-	1.5	-
Small wood remains *	3.7	3.4	3.7	6.4	0.5	1.3	-	7.0	0.1	-	-
Fraction size <0.25 mm *	16.1	3.6	36.1	42.2	99.3	80.0	52.2	142	25.0	80.7	36.1
Total *	20.0	7.4	40.5	50.0	107	83.6	81.6	173	27.3	153	93.6

AT – the upper layer of paleosol with traces of peat accumulation. A1 – the upper layer of paleosol with raw humus. ATC - transient layer between horizons of peat accumulation and underlying parental mineral matter with several meters in thickness sometimes. AC - transient layer between horizon with raw humus and underlying mineral matter. ATB - transient layer between horizon of peat accumulation and underlying horizon.

Seeds of *Ranunculus*, with dominance of *Ranunculus repens,* were often found mixed with buds of *Bistorta vivipara*. *Ranunculus* seeds were relatively well -preserved with natural colors and easily noticeable against a dark background of *Bistorta vivipara* buds.

Many seeds were empty inside. This might be due to the collection in the seeds milky stage and/or the destructive activity of saprophytic microflora. Seeds were often penetrated and conglomerated with recently dead fungal mycelia, and/or with different viable microorganisms [15] which caused seeds, when improperly dried at room temperature, to mold in a few days.

Table 3. Plant remains from soils of cryopedolith layers
(pieces and $g*$ remains of soil sample unit)

Species	Sampling layers					
	(a)	(b)	(c)	(d) (L)	(e) (M)	(f) (U)
Moss, Aulacomnium palustre	–	–	–	–	–	1
Moss, Aulacomnium turgidum	–	–	–	–	–	2
Arctous alpine (seeds)	17	–	1	6	87	–
Astragalus alpinus (seeds)	–	–	–	–	1	–
Bistorta elliptica (seeds)	3	–	1	–	–	–
Carex appendiculata (seeds)	–	38	301	–	–	–
C. aterrima (seeds)	–	–	3	–	–	–
C. concolor(seeds)	10	12	60	6	–	2
C. juncella (seeds)	–	2	–	–	–	–
C. lugens (seeds)	–	–	–	8	3	–
C. rhynchophysa (seeds)	–	–	–	–	1	–
Carex sp. (roots)*	–	–	0.4	–	–	–
Comarum palustre (bark)*	0.01					
Draba nemorosa (seeds)	–	–	–	–	3	–
Dryas octopetala subincisa (leaf)	–	–	–	–	1	–
Empetrum androgynum (leaf)	–	–	–	–	–	1
Equisetum pratense	–	–	–	–	–	1
Larix cajanderi (bark,leaf)*	–	–	0.2	–	–	0.01
L. cajanderi,(seeds)	–	–	–	–	–	–
Ledum decumbens,(leaves)	–	–	–	–	–	1
Potentilla stipularis (seeds)	11	–	–	3	54	1
Puccinella sp.(seeds)	1	–	–	–	–	–
Kobresia myosuroides,(seeds)	–	1	–	–	–	–
Salix sp.(bark)*	0.2	0.2	1.5	–	–	–
Silene stenophylla,(seeds)	–	–	–	–	1	–
Tanacetum vulgare,(seeds)	–	–	–	–	–	1
Wood of shrubs*	–	–	–	0.2	1.0	0.3
Vegetative remains*	0.5	7.1	66.6	2.6	6.0	0.5
Fraction (size < 0.25 mm*	82.6	99.3	80.7	11.1	40.8	27.9
Total weight (g)*	83.1	107	153	13.9	47.8	28.7

Sampling layers: (a) P-1212,1; (b) F1-1321; (c) P-01-03; (d) P-8-09; (e) P-7-09; (f) P-6-09. (L) lower, (M) middle, (U) upper.

All the species identified belong to different ecotopes of wooded tundra of open tree formations on uplands; river floodplains and other water-logged areas; unmated patches of disturbed lands; and steppificated cleves. A number of these plants inhabit both tundra and wooded tundra, but some of them can only be found in tundra, for example: *Vaccinium minus, Papaver polare, Puccinella borealis, Dryas octopetala* subsp. *subincisa, Salix polaris.*

PLANT REMAINS IN CRYOPEDOLITH AND PALEOSOLS

Ancient plant seeds and remains were also found in cryopedolith, or within profiles of buried epigenetic soils [5, 23]. In paleosols, formed during the warm spell between 37 – 26 ka, fossilized plant seeds were found only in the uppermost layers. They were not well preserved in comparison to seeds buried in ancient rodent burrows. They were broken into pieces, seed surface characteristics wiped off the surface with traces of microbial destruction, etc. These sometimes made it difficult or even impossible to identify to species. In such cases, taxonomic identification was successful only through tribe or *genus*. Some *genera* were identified easily as (a) willows (*Salix*) based on distinctive features of the leaves, cortex and fruitcase; (b) some monocots, especially of *Poaceae*, based on morphological characteristics of the leaves, culms and the absence of panicle remains; (c) and species of *Rubus chamaemorus* by fragments of stalk; *Empetrum androgynum* by fragments of stalk and leaves; *Chamaenerion angustifolium,* and *Ch. latifolium* by fragments of leaves and stems; *Equisetum* by remains of vegetative parts; *Dryas* by leaves; *Larix* by leaves and leaf-spurs; *Duschekia fruticosa* by well-preserved anthers (Table 2). A major part of organic remains in cryo-pedolith was represented by thin vegetative detritus consisting of 0.1 – 0.01 mm (in diameter) fragments of roots and vegetative parts.

Seeds of the following plant species were identified in the soil samples from the uppermost layers of these paleosols: *Arctagrostis latifolia, Arctous alpina, Aconogonon tripterocarpum, Aulacomnium palustre, A. p. var. imbricatum, Betula divaricata, Bistorta elliptica, Carex aterrima, Carex concolor, Carex rhynchophysa, Comarum palustre, Larix cajanderi, Potentilla stipularis* and *Kobresia myosuroides*. Seeds in cryopedolith occurred very rarely, yet they were often better preserved then those of in deeper soils, probably due to the smaller number of saprophytic microorganisms (Table 3).

CONCLUSIONS

Fossilized seeds and plant remains of 40 taxa, collected from the permafrost and ancient rodent burrows of the chapter presented, provides uniquely preserved genetic material for further archaeogenetical studies to track crop and weed microevolution and domestication.

ACKNOWLEDGMENT

Authors wish to thank *Mrs. Elena Schukina* for the comments on the manuscript.

REFERENCES

Bolikhovskaya NS (1995) *Evolution of the Loess-soil formation of Northern Eurasia Mosk.* Gos. Univ., Moscow 270 pp. (in Russian).

Cherepanov SK (1995) *Vascular Plants of Russia* (the former USSR). North American Branch, Cambridge University Press, 991 p.

Gubin SV (1998) *Soil Formation during the Sartan Cryochron in the Western sector of Beringia.* Pochvovedenie, N 5, 1998, c. 605-609.(in Russian).

Gubin SV and Khasanov BF (1996) Fossil Burrows of Mammals in the Loess-Ice Deposits of the Kolyma-Indigirka Lowland, Doklady AN, vol. 346, no. 2, pp. 278–279 [*Doklady Biol. Sci.* (Engl. Transl.), vol. 346, pp. 26–27].

Gubin SV (1984) Paleopedological Analysis of Late Pleistocene (Yedoma) Deposits in the Duvanny Yar Exposure. *Bull. of Quaternary commission,* 53, 125-128. (in Russian).

Gubin SV (2002) Pedogenesis as Part of the Mechanism of Deposit Formation in Late Pleistocene ice complex. Kriosfera Zemli [*Earth Cryosphere*] Vol.VI № 3:82–91 (in Russian).

Gubin SV, Zanina OG, Maksimovich SV, Kuzmina SA, ZazhiginVS (2003) Reconstruction of Ice Complex Sediment Formation Conditions based on the Results of Research of Late Pleistocene Rodent Burrows. *The Earth Cryosphere*, 7, 3, 13-22 (in Russian).

Başli AG, G Gyulai, Z Tóth, A Güner, Z Szabó, VL Stakhov, L Murenyetz, SG Yashina, L Heszky, SV Gubin (2009) Light and Scanning Electron Microscopic Analysis of *Silene stenophylla* Seeds Excavated from Pleistocene-Age (Kolyma). *Anadolu Univ J Sci and Technol* 10:161-167.

Kaplina TN, Giterman RE, Lakhtina OV, Abrashov BA, Kiselyov SV, Sher AV (1978) Duvanny Cliff as the Key Section of the Late Pleistocene Deposits of the Kolyma Lowland. *Bulletin of Quaternary Commission* 48, 49-65 (in Russian).

Morozova TD (1971) On Characteristics of Loess Deposits in Central Yakutia. *Izvestiya AN USSR, Ser Geogr.*, №5 (in Russian).

Rivkina E, Gilichinskiy D, Wagener S, Tiedje J, McCrath J (1998) Biogeochemical Activity of Anaerobic Microorganisms from Buried Permafrost Sediments, *Geomicrobiology*, No. 15, pp. 187–193.

Sher A and Kuzmina S (2007) Beetle records/Late Pleistocene of Northern Asia. In: Elias, S. (Ed.), Encyclopaedia *of Quaternary Science*, vol. 1. Elsevier, Amsterdam, pp. 246–267.

Sher AV, Kaplina TN, Ovander MG (1987) Unified Regional Stratigraphic Chart for the Quaternary deposits in the Yana-Kolyma Lowland and its Mountainous Surroundings. In. *Explanatory Note—Decisions of Interdepartmental Stratigraphic Commission on the Quaternary of the Eastern USSR.* SAS FEB Publishers, North-Eastern Complex Research Institute Magadan, pp. 29–69 (in Russian).

Sher AV, Kuzmina SA, Kuznetsova TV, Sulerzhinsky LD (2005) New Insights into the Weichselian Environment and Climate of the East Siberian Arctic Derived from Fossil Insects, Plants, and Mammals. *Quaternary Science Reviews,* 24, pp. 533–569.

Stakhov VL, Gubin SV, Maksimovich SV, Rebrikov DV, Savilova AM, Kochkina GA, Ozerskaya SM, Ivanushkina NE, Vorobyova EA (2008) Microbial Communities of Ancient Seeds Derived from Permanently Frozen Pleistocene Deposits, *Microbiology*, Vol. 77, No. 3, pp. 348–355.

Tomirdiaro SV (1980) Loess-Ice Formation of the East Siberia in the Late Pleistocene and Holocene. *Nauka*, Moscow (in Russian).

Vinogradov BS (1948) On Discovery of Fossil Ground Squirrel Corpses in Permafrost Indigirka River Basin. *Doklady AN USSR.* V 62. № 4. C.337-353. (in Russian).

Vishnivetskaya TA, Erokhina LG, Gilichinskii DA, Vorobyova EV (1997) Blue–Green Algae from Arctic Permafrost Sedimentary Rocks, *Kriosfera Zemli*, no. 2, pp. 71–76.

Vorobyova E, Soina V, Gorlenko M, Minkovskaya N, Zalinova N, Mamukelashvili A, Gilichinskiy D, Rivkina E, Vishnivetskaya T (1997) The Deep Cold Biosphere: Facts and Hypothesis, *FEMS. Microbiol. Rev.*, no. 20, pp. 227–290.

Yashina SG, Shabaeva EV, Gubin SV, Maksimovich SV (1998) Viable Cells of Plant Seeds Isolated from Permafrost Sediments of Northern Yakutia: Conservation of Genetic Resources. *Proc. XV Workshop, Pushchino*, pp. 229–231 (in Russian).

Yashina SG, Gubin SV, Shabaeva EV, Egorova EF, Maksimovich SV (2002) Viability of Higher Plant Seeds of the Late Pleistocene Age from Permafrost Deposits as Determined by in vitro Culturing. *Doklady Biological Sciences*, Vol. 383, pp. 151–154.

Zanina OG (2005) Fossil Rodent Burrows from Frozen Late Pleistocene Deposits of the Kolyma Lowland, *Zool. Zh.* Vol. 84, no. 6, pp. 728–736 (in Russian).

Zanina OG (2006) Pedogenesis and Environment during Karginsky Time on Kolyma Lowland. PhD *Dissertation in Geography*. IG RAS, Moscow.

Zazula GD, Froese DG, Schweger CE, Mathewes RW, Beaudoin AB, Telka AM, Harington CR, Westgate JA (2003) Ice age steppe vegetation in East Beringia. *Nature* 423, 603.

Zazula GD, Froese DG, Westgate JA, Lafarge C, Mathewes RW (2005) Paleoecology of Beringian "Packrat" Middens from Central Yukon Territory. *Quaternary Research* 63:189–198.

Zvyagintsev DG, Gilichinsky DA, Blagodatskii SA, Vorobyova EA, Khlebnikova GM, Arkhangelova AA, Kudryavtseva NN (1985) Duration of Microbial Survival in Permafrost Sedimentary Rocks and Buried Soils, *Mikrobiologiya*, vol. 54, no. 1, pp. 155–161.

In: Plant Archaeogenetics
Editor: Gábor Gyulai, pp. 23-30

ISBN: 978-1-61122-644-7
© 2011 Nova Science Publishers, Inc.

Chapter 3

MORPHOGENETICS OF DICOTS AND LARGE- AND SMALL-GRAINED WILD GRASSES FROM THE PALEOLITHIC ERA (OLD STONE AGE) OHALO II., ISRAEL (23,000 BP)

Ehud Weiss[1,2,], Mordechai E. Kislev[3], Orit Simchoni[3] and Dani Nadel[4]*

[1]Martin (Szusz) Department of Land of Israel Studies and Archaeology, The Institute of Archaeology, Bar-Ilan University, Ramat Gan 52900, Israel
[2]Kimmel Center for Archaeological Sciences, Weizmann Institute of Science, Rehovot 76100, Israel
[3]Faculty of Life Sciences, Bar-Ilan University, Ramat Gan 52900, Israel
[4]Zinman Institute of Archaeology, University of Haifa, Haifa 31905, Israel

ABSTRACT

Ohalo II is a submerged, late Upper Paleolithic (locally termed Early Epipalaeolithic) site, radiocarbon dated to ca 22,500–23,500 BP. (Kislev *et al.* 1992). The site is located on the southwestern shore of the Sea of Galilee (Lake Kinneret), Rift Valley, Israel (Figure 1). This hunter-gatherer-fisher camp covers more than 2000m^2 (0.2 ha) and includes the remains of six brush huts with numerous plant remains, open-air hearths, and a human grave. The site was occupied during the Last Glacial Maximum (LGM), a period of cold and dry climate when ice sheets covered parts of North America and Europe. Almost 100 taxa are represented among the identified plant remains (Nadel 1997; Simchoni 1998; Weiss 2009). The most numerous monocot and dicot species will be discussed in this Chapter.

---

* E-mail: eweiss@mail.biu.ac.il

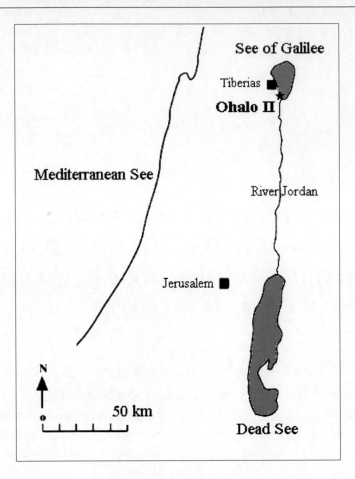

Figure 1. Location of Ohalo II., the late Upper Paleolithic site with hunter-gatherer-fisher's camp from ca. 22,500–23,500 BP.

INTRODUCTION

Ohalo II is an exceptional site on a number of accounts. No other Levantine Upper Paleolithic site has produced such well-preserved and large quantities of archaeological materials. These include rich plant assemblage, flint and ground stone tools, and a broad spectrum of animal remains (mammals, birds, rodents, fish, mollusks) (Kislev *et al.* 1992; Simmons and Nadel 1998; Nadel and Werker 1999; Belmaker *et al.* 2001; Nadel *et al.* 1994, 2002; Rabinovich 2002; Rabinovich and Nadel 1994, 2002; Simmons 2002; Weiss 2002, 2009; Weiss *et al.* 2008).

Three successive seasons (1989–1991) of excavations at Ohalo II were conducted immediately following its discovery, when the lake level had dropped and much of the site was exposed (Figure 1). Subsequently, the site was submerged again for several years, rendering fieldwork difficult. However, three additional seasons of fieldwork took place in 1999–2001, after droughts and heavy water pumping from the lake that dropped the water level to ca. 213.80 m below msl.

The archeological remains are located on the clay/silt Lisan Formation bedrock. The Lisan Formation into which the Ohalo II hut floors were dug is composed of lacustrine sediments, dated between 70,000 and 15,000 years BP (Bartov *et al.* 2002). The huts were located on the shores of the newly formed Sea of Galilee (Belitzky and Nadel 2002).

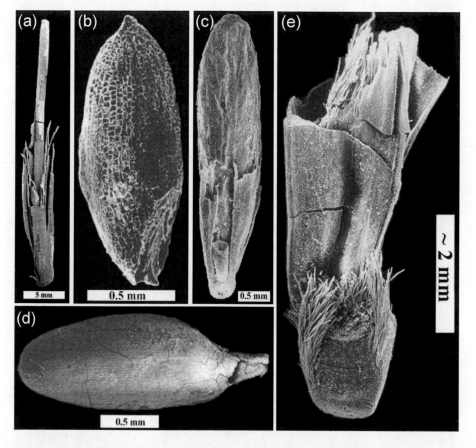

Figure 2. SEM micrograph of 22,500–23,500 year-old monocots excavated at Ohalo II., Israel. *(a)* Shoot of alkaligrass (*Puccinellia cf. convoluta*) with the basal part of the flowering stem crowded by open leaf sheaths (scale bar 1mm) (Nadel *et al.* 2004). *(b)* Grain of alkaligrass (*Puccinellia cf. convoluta*); the cubical epidermis cells, and on the bottom-right side the embryo and its radicle can be seen (Simonchi 1998; Weiss 2004). *(c)* Grain of brome (*Bromus pseudobrachystachys / tigridis*); the basal parts of the lemmas are preserved, on the bottom the floret node and the arched rachilla joint (Simonchi 1998; Weiss 2004). *(d)* Grain of bladder/creeping foxtail (*Alopecurus utriculatus / arundinaceus*); in the right, the grain's elongated, tipped base holding the embryo on its dorsal side (Simonchi 1998; Weiss 2004). *(e)* Dispersal unit of wild wheat (*Triticum dicoccoides*), ventral view. Notice the delicate hairs on the node, the grain that enclosed inside the unit, with hairs on top of it as well (Weiss 2009).

The probable reason for the outstanding preservation is the rise of the lake level, which sealed the site immediately after its abandonment (Nadel *et al.* 2004) similar to well-preserved samples due to anaerobic conditions (Lágler *et al.* 2005; Szabó *et al.* 2005; Gyulai *et al.* 2006). The plant material, in particular, was protected by two successive events. The first was charring by fire, which preserved most of the archaeobotanical remains. Then, very shortly after the site was abandoned, the water level of the newly formed Lake Galilee rose. This combination of charring, evidently in low-oxygen conditions (perhaps under the cover of the hut's walls), and sealing under lake sediments and water apparently created ideal conditions for preservation of organic materials (Weiss 2002). The vast majority of the site's plant remains; all but 152 seeds were charred (Kislev *et al.* 1992, 2002; Simchoni 1998; Weiss 2002).

Figure 3. SEM micrograph of ca. 22,500–23,500 year-old dicots excavated at Ohalo II., Israel. *(a and b)* Groundsel (*Senecio glaucus*), the head is seen from the outside. It has five upper whorl bracts; the one on the left is broken. Attached to these bracts is another, thinner and pointed bract from a lower whorl. *(b)* The outside of the head in the foreground with three of the originally five achenes and bracts behind them. The achenes are covered with short hairs; the pappus' attachment scar can be seen on their top. The depressions in front of the achenes indicate where the missing achenes were attached to the head (Weiss 2002). *(c)* Pip of wild grape (*Vitis sylvestris*), ventral view, the beak is short and blunt (Kislev *et al.* 1992). *(d)* Tabor oak (*Quercus ithaburensis*) premature cup, side view. The tiny bracts of this 1.5 mm diameter cup were preserved, folded backward. The upper part of the acorn – that protrude over the cup – is broken (Weiss 2009). *(e)* Egyptian sea-blite (*Suaeda aegyptiaca*) flower, side and half-up view. From the original five perianth leaves, three preserved, holding the stamens' filament at their base. In the middle of the flower – an elongated pear-like ovary, with a shallow depression on its top. A very short style is emerging from the bottom of the depression and split into three stigmas (Weiss 2009).

Table 1. Representative plant assemblage from the huts Ohalo II (Weiss *et al.* 2008)

#	Taxon (organ)	Quantity
1	*Adonis dentata/microcarpa* (mericarp)	71
2	*Apiaceae* (*Umbelliferae*) (umbel and umbellule)	253
3	*Atriplex rosea/leucoclada* (fruit and seed)	1228
4	*Bromus pseudobrachystachys / tigridis* (grain)	9904
5	*Hordeum marinum/hystrix* (grain)	505
6	*Hordeum spontaneum* (grain)	606
7	*Malva parviflora* (mericarp)	594
8	*Melilotus indicus* (seed)	134
9	*Piptatherum holciforme* (grain)	819
10	*Puccinellia cf. convoluta* (grain)	673
11	*Rubus sanguineus/canescens* (nutlet)	175
12	*Silybum marianum* (achene)	115
13	*Suaeda palaestina / fruticosa* (seed)	28,373

MATRIALS AND METHODS

The botanical remains were distinguished by exquisite preservation. As a result, they were identified to the family, genus, and in many cases, even to the species level (Table 1). Seeds were identified by morphogenetic characters (Katz *et al.* 1965; Beijerinck 1976; Feinbrun-Dothan 1978, 1986; Zohary 1966, 1972).

CONCLUSIONS

Due to the high number of plant remains identified to the species level, we were able to reconstruct the ancient ecosystem and landscape around the site. The park-like forest on the slopes of the eastern Lower Galilee and western Golan Heights had trees like Christ's thorn (*Ziziphus lotus*) and *Z. spina-christi*), hawthorn (*Crataegus aronia* and *C. azarolus*), Mount Tabor oak (*Quercus ithaburensis*), officinal storax (*Styrax officinalis*), wild pistachio (*Pistacia atlantica*), wild almond (*Amygdalus communis* and *A. korschinskii*), and wild olive (*Olea europaea*). The grass species of this habitat included foxtail (*Alopecurus arundinaceus* and *utriculatus*), wild barley (*H. spontaneum*), and wild emmer wheat (*T. dicoccoides*).

From the variable environment around Ohalo II, the site inhabitants gathered vegetal food. Wild barley and wild emmer wheat, together with acorns of Mount Tabor oak, provided the major carbohydrates. Fruits and berries, like Christ's thorn, wild fig, and wild grape, provided sugars. Almond, pistachio, and wild olive supplied the fats. Although legumes like lentil, pea, and vetch are generally a good source of proteins, very few pulses were found at Ohalo II. The reason for the lack of pulses cannot be attributed to preservation, in light of the high level of preservation of other finds. The location of the site next to the lake and the park-like forest may explain the large amount of animal bones, such as fish, terrestrial mammals, and birds. The Ohalo II inhabitants apparently relied heavily on animal protein for their diet.

The small grained grasses of the four taxa *Bromus pseudobrachystachys / tigridis* (Figure 2a), *Hordeum marinum/hystrix, H. spontaneum, Puccinellia cf. convoluta* (Fig 2b) were part of the diet of the site's inhabitants (Weiss *et al*. 2005), and their grains most probably were ground on the grinding stone or processed in other ways near it (Piperno *et al*. 2004). *Alopecurus utriculatus / arundinaceus* (Figure 2c) and *Aegilops* spp. (including *Ae. peregrina* and *Ae .geniculata / peregrina*) grains were also found.

Holy thistle (*Silybum marianum*) leaves and stalks are edible (extracts of *S. marianum* achenes are a common alternative medicine today) (Table 1). Mallow (*Malva parviflora*) leaves and fruits might be also a food plant. It is likely that *Atriplex* branches were used in the construction of the huts. *Rubus* fruits are edible and nutlets of several *Rubus* species have been found at many European sites, from the Neolithic period onward (Renfrew 1973; Zohary and Hopf 1994). It is most likely that these fruits were eaten by the Ohalo II dwellers. Due to the very high lignin content in their seed coating, some archaeological *Rubus* seeds were preserved in an uncharred state (G. Hillman, personal communications). *Adonis dentate / microcarpa* (peasant's eye) fruits are spike-like, dense, and apocarpic, bearing many mericarps. All the umbels and umbellules of *Apiaceae* were similar and apparently belonged to the same species. *Apiaceae* were not the only family of flowers represented in the Ohalo II botanical assemblage. *Senecio glaucus* (groundsel) capitula (heads) (Compositae/Asteraceae) were also found (Figure 2d). In total, 20 Compositae heads were found. It is likely that the tens of thousands of *Suaeda palaestina / fruticosa* seeds came from the branches that formed the walls of the huts (Nadel and Werker 1999).

To conclude, morphogenetics of the seeds, grains and plant remains excavated at Ohalo II before the Neolithic Revolution provide unique materials for a contemporary analysis tracing plant evolution and domestication.

REFERENCES

Bartov Y, Stein M, Enzel Y, Agnon A, Reches Z (2002) Lake levels and sequence stratigraphy of Lake Lisan, the Late Pleistocene precursor of the Dead Sea. *Quaternary Research* 57: 9–21.

Belitzky S and Nadel D (2002) The Ohalo II prehistoric camp (19.5 Ky): new evidence for environmental and tectonic changes at the Sea of Galilee. *Geoarchaeology: An International Journal* 17: 453–464.

Belmaker M, Nadel D, Tchernov E (2001) Micromammal taphonomy in the site of Ohalo II (19 ky., Jordan Valley). *Archaeofauna* 10: 125–135.

Beijerinck W (1976) Zadenatlas der Nederlandsche Flora: Ten Behoeve van de Botanie, Palaeontologie, Bodemcultuur en Waren-Kennis. Amsterdam.

Feinbrun-Dothan N (1978) *Flora Palaestina, Part 3*. Jerusalem.

Feinbrun-Dothan N (1986) *Flora Palaestina, Part 4*. Jerusalem.

Gyulai G, Humphreys M, Lágler R, Szabó Z, Tóth Z, Bittsánszky A, Gyulai F, Heszky L (2006) Seed remains of common millet from the 4th (Mongolia) and 15th (Hungary) centuries: AFLP, SSR and mtDNA sequence recoveries. *Seed Science Research* 16: 179-191.

Katz NJ, Katz SV, Kipiani MG (1965) Atlas i opredelitelj plodov i semjan, v strecajushija v cetverticnjih otlozenijah SSSR (Atlas and Keys of Fruits and Seeds Occurring in the Quaternary Deposits of the USSR). Moscow.

Kislev ME, Nadel D, Carmi I (1992) Epipalaeolithic (19,000 B.P.) cereal and fruit diet at Ohalo II. Sea of Galilee. *Review of Palaeobotany and Palynology* 73:161–166.

Lágler R, Gyulai G, Humphreys M, Szabó Z, Horváth L, Bittsánszky A, Kiss J, Holly L, Heszky L (2005) Morphological and molecular analysis of common millet (*P. miliaceum*) cultivars compared to an aDNA sample from the 15th century (Hungary). *Euphytica* 146: 77-85.

Nadel D (1997) The spatial organization of prehistoric sites in the Jordan Valley: Kebaran, Natufian and Neolithic case studies, *Ph.D. Thesis,* Institute of Archaeology, The Hebrew University, Jerusalem.

Nadel D and Werker E (1999) The oldest brush hut plant remains from Ohalo II, Jordan Valley, Israel (19 ka B.P.). *Antiquity* 73: 755–764.

Nadel D, Danin A, Werker E, Schick T, Kislev ME, Stewart K (1994) 19,000-year-old twisted fibers from Ohalo II. *Current Anthropology* 35: 451–458.

Nadel D, Tsatskin A, Bar-Yosef Mayer DE, Belmaker M, Boaretto E, Kislev ME, Hershkovitz I, Rabinovich R, Simmons T, Weiss E, Zohar I, Asfur O, Emmer G, Ghraieb T, Grinberg U, Halabi H, Weissbrod L, Zaidner Y (2002) The Ohalo II 1999–2000 seasons of excavation: a preliminary report, Mitekufat Haeven. *Journal of the Israel Prehistoric Society* 32: 17–48.

Nadel D, Weiss E, Simchoni O, Tsatskin A, Danin A, Kislev M (2004) Stone Age hut in Israel yields world's oldest evidence of bedding. *Proceedings of the National Academy of Sciences of the United States of America* 101: 6821–6826.

Piperno DR, E Weiss, I Holst, D Nadel (2004) Procesing of wild cereal grains in the Upper Palaeolithic revealed by starch grain analysis. *Nature* 430: 670-673.

Rabinovich R (2002) The mammal bones: environment, food and tools. In: Nadel, D. (Ed.), Ohalo IIdA 23,000-Year-Old Fisher-Hunter-Gatherers' Camp on the Shore of the Sea of Galilee, Hecht Museum. University of Haifa, Haifa, pp. *24–*27. Catalogue No. 20pp.

Rabinovich R and Nadel D (1994) Bone tools from Ohalo IIda morphological and functional study, Mitekufat Haeven. *Journal of Israel Prehistoric Society* 26: 32–63.

Rabinovich R and Nadel D (2002) Broken mammal bones: taphonomy and food sharing at the Ohalo II submerged prehistoric camp, Archaeozoology of the Near East. ARC-Publicaties 62, Groningen Six symposium, The Netherlands.

Renfew JM (1973) *Palaeoethnobotany, the Prehistoric Food Plants of the Near East and Europe.* Columbia University Press, New York.

Simchoni O (1998) Reconstruction of the Landscape and Human Economy 19,000 B.P. in the Upper Jordan Valley by the Botanical Remains Found at Ohalo II, *Ph.D. Thesis*, Faculty of Life Sciences, Bar-Ilan University, Ramat Gan.

Simmons T (2002) *The birds from Ohalo II. In: Nadel, D. (Ed.), Ohalo IId A 23,000-Year-Old Fisher-Hunter-Gatherers' Camp on the Shore of the Sea of Galilee,* Hecht Museum. University of Haifa, Haifa, pp. 32–36. Catalogue No. 20pp.

Simmons T and Nadel D (1998) The avifauna of the early Epipalaeolithic site of Ohalo II (19,400 years B.P.), Israel: Species diversity, habitat, and seasonality. *International Journal of Osteoarchaeology* 8: 79–96.

Szabó Z, Gyulai G, Humphreys M, Horváth L, Bittsánszky A, Lágler R, Heszky L (2005) Genetic variation of melon (*C. melo*) compared to an extinct landrace from the Middle Ages (Hungary) I. rDNA, SSR and SNP analysis of 47 cultivars. *Euphytica* 146: 87-94.

Weiss E (2002) Issues in reconstruction of the human economy and society of the Epipalaeolithic site Ohalo II inhabitants by the macrofossil botanical remains, *Ph.D. Thesis,* Department of Land of Israel Studies and Archaeology, Bar-Ilan University, Ramat-Gan.

Weiss E (2009) Glimpsing into a hut: The economy and society of Ohalo II's inhabitants. In: *From forages to farmers*: Gordon C Hillman Festschrift. Eds. AS Fairbairn and E Weiss. Oxbow Books, Oxford and Oakwille. pp. 153-160. ISBN 1842173545.

Weiss E, Kislev ME, Simchoni O, Nadel D (2004) Small-grained wild grasses as staple food at the 23,000 year old site of Ohalo II, Israel. *Economic Botany* 58(Supplement): S125–S134. 2004 125–134.

Weiss E, Kislev ME, Simchoni O, Nadel D, Tschauner H (2008) Plant-food preparation area on an Upper Paleolithic brush hut floor at Ohalo II, Israel. *Journal of Archaeological Science* 35:2400–2414.

Zohary M (1966) *Flora Palaestina,* Part 1. Jerusalem.

Zohary M (1972) *Flora Palaestina,* Part 2. Jerusalem.

Zohary D and Hopf M (1994) *Domestication of Plants in the Old World,* second ed. Oxford Science Publications, Oxford.

In: Plant Archaeogenetics
Editor: Gábor Gyulai, pp. 31-39

ISBN: 978-1-61122-644-7
© 2011 Nova Science Publishers, Inc.

Chapter 4

MORPHOGENETICS OF SEEDS AND PLANT REMAINS IN THE CARPATHIAN BASIN FROM THE NEOLITHIC TO THE LATE MEDIEVAL AGE (8000 BP TO 17TH CENT CE)

Ferenc Gyulai[1]*, Árpád Kenéz[1] and Ákos Pető[2]

[1]Institute of Environmental Sciences, St. Stephanus University, Gödöllő, 2103, Hungary
[2]Centre for Cultural Heritage, Hungarian National Museum,
13-17 Dugovics T sqr, Budapest 1036, Hungary

ABSTRACT

Morphogenetics of plant remains excavated in the Carpathian Basin from 8000 BP (*before present*) to the late Medieval Age (17th cent. CE) are discussed in this chapter. The collection of samples goes back more than 130 years (1877 – 2010), consisting of 9 million seeds of 660 plant taxa from 460 archaeological sites representing the results of 24 researchers. As the Carpathian Basin is one of the longest continuously inhabited regions in Europe, the morphogenetical analysis of the most ancient seeds helps to trace archaeogenetical events, the origin of agriculture, weed history and environmental changes.

INTRODUCTION

The first domesticated forms of wheat and barley appeared around 8000-7000 BCE (*before current era*) in the Middle East during the "Neolithic Revolution" (Zohary and Hopf 1988). Plant domestication in Europe started around 6-8 thousand years ago, and included the Carpathian Basin (the historical Hungary), with one of the most ancient Neolithic agrarian

---

* E-mail: gyulai.ferenc@kti.szie.hu

Körös-Starčevo culture in Europe The cultivation of vegetables has four thousand years and fruit cultivation at least two thousand years of history in this region (Gyulai 2000).

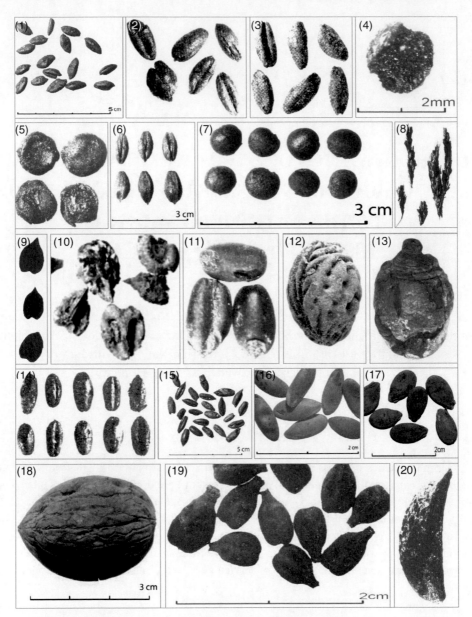

Figure 1. Samples of ancient seed remains excavated in the Carpatian Basin. (1) Einkorn (*Triticum monococcum*) naked grains from Dévaványa-Réhelyi, *Middle Neolithic Tisza Culture* (6000 BCE), (Source: Hungarian Agricultural Museum, Budapest). (2) Emmer (*Triticum turgidum* subsp. *dicoccum*) grains from Dévaványa-Réhelyi (6000 BCE) (Hartyányi and Máthé 1980) and (3) from Berettyóújfalu-Szilhalom (5000 BCE) (Hartyányi and Máthé 1980). (4) Millet (*Panicum miliaceum*) naked grain from Ibrány, *Early Neolithic Körös Culture* (beginning of the 6000 BCE), (Photo by Á Kenéz). (5) Field pea (*Pisum sativum* var. *arvense*) seeds from the *Late Neolithic* settlement of Berettyóújfalu-Szilhalom (5000 BCE) (Source: Hartyányi and Máthé 1980). (6) Six-rowed barley (*Hordeum vulgare* subsp. *hexastichum*) hulled grains from Alpár-Várdomb, *Vatya Culture* (17th–16th CENT. BCE) (Source: Hungarian Agricultural Museum, Budapest). (7) Small seed lentil (*Lens culinaris* subsp. *microsperma*)

from Alpár-Várdomb, *Vatya Culture* (17th–16th CENT. BCE) (Source: Hungarian Agricultural Museum, Budapest). (8) Charcoaled cereal spikes from a *Roman watchtower*, Budakalász-Őrtorony, Lupa csárda (3rd–4th CENT. CE) (Source: Hungarian Agricultural Museum, Budapest. (9) Grape (*Vitis vinifera*) from Roman Age, Kaszásdűlő-Raktárrét (2nd – 3rd CENT.CE) (computer image) and (10) late Roman Age, Budapest, Vörösvári str. 20 (3rd–4th cent. CE) (Source: Hungarian Agricultural Museum, Budapest). (11) Bread wheat (*Triticum aestivum* subsp. *vulgare*) grains from Fonyód-Bélatelep, Late Migration Period lake dwelling settlement (7th–9th CENT. CE) (Photo by F Gyulai). (12) Peach (*Prunus persica*) stone from Fonyód-Bélatelep, *Late Migration Period* settlement (9th CENT. CE) (Photo by F Gyulai). (13) Apricot (*Prunus armeniaca*) fruit mummy with *Monilia* damage from Fonyód-Bélatelep, *Late Migration Period* settlement (9th CENT. CE) (Photo by F Gyulai). (14) Bread wheat (*Triticum aestivum* subsp. *vulgare*) grains from Sopron –Városháza u (9th–10th CENT. CE) (Hartyányi *et al.* 1967–68). (15) Rye (*Secale cereale*) grains from the ispán's castle at Hont, *Early Árpádian Period* (10th CENT. CE) (Source: Hungarian Agricultural Museum, Budapest. Seeds of (16) melon (*Cucumis melo*), (17) watermelon (*Citrullus lanatus*), (18) walnut (*Juglans regia*) and (19) grapes (*Vitis vinifera* subsp. *vinifera*) from well #15 of the former Military Headquarters in the Buda castle (15th CENT. CE) (Photos by F Hegyi). (20) Garlic (*Allium sativum*) clove from Nagyvázsony-Csepely (15th–16th CENT. CE) (Source: Hungarian Agricultural Museum, Budapest).

MATERIALS AND METHODS

For seed morphogenetical characterization, handbooks were used (Beijerinck 1947; Brecher 1960; Schermann 1966; Schoch *et al.* 1988). For computer image analysis, morphometry was applied according to Rovner and Gyulai (2007), and for stratigraphic bar diagram, C2 program was used (Juggins 2007).

CONCLUSIONS

Early Neolithic (> 6000 - 5300 BCE): Körös-Starčevo Culture

The Körös-Starčevo culture goes back 8000 BP with intensive farming, animal husbandry and fishing culture along the rivers, Tisza and Körös of the Hungarian Great Plain (the *Puszta*), mainly in areas with sandy soils. Excavated seeds showed the dominance of hulled wheats of einkorn (*Triticum monococcum* subsp. *monococcum*) and emmer (*T. turgidum* subsp. *dicoccum*), as well as the six-rowed (*Hordeum vulgare* subsp. *hexastichum*) and two-rowed barley (*H. v.* subsp. *distichun*) with naked and hulled grains. The earliest known occurrence of bread wheat (*Triticum aestivum* subsp. *vulgare*), club wheat (*T. ae.* subsp. *compactum*), spelt (*T. ae.* subsp. *spelta*), millet (*Panicum miliaceum*), and oil and fiber plants of poppy (*Papaver somniferum*) and flax (*Linum usitatissimum*) were also from this period (Figures 1, 2).

Middle (5300-4700 BCE) and Late Neolithic (4700-4300 BCE): Linear Pottery Culture

The Linear Pottery Culture of the Middle Neolithic Age along the river Tisza was found mostly on the fertile loess soil, which indicates that inhabitants recognized loess fertility first.

Great numbers seeds of six-rowed barley, einkorn, emmer, and vegetables including lentils (*Lens culanaris* subsp. *microsperma*), common vetch (*Vicia sativa*) and peas (*Pisum sativum* subsp. *arvense*) were excavated from these sites (Figures 1, 2).

Early (4300-3500 BCE) and Late Copper Age (3500-3000 BCE)

By the end of the Neolithic Age, the warm and mild humid climate gradually cooled down. The newly settled human populations started large scale deforestation in order to clear land for the agriculture under the cool and more humid climate.

Early - (3,000-1,600 BCE), Middle - (1600-1300 BCE) and Late Bronze Age (1300-900 BCE)

Ceramics from around 3000 BCE indicate that new groups arrived in the Carpathian Basin from the east, together with a mediterranean population from the south, during this period (Gyulai 1993). Excavation sites show remains of intensive animal breeding and plant cultivation. The most important cereals were barley, einkorn and emmer, with legumes including lentils, peas, chick peas (*Cicer arietinum*), bitter vetch (*Vicia ervilia*), horse bean (*Vicia faba*), and fiber plants of flax (*Linum usitatissimum*) and gold-of-pleasure (*Camelina sativa*) (Figures 1, 2). By the end of the Bronze Age, around 1300 BCE, the Carpathian Basin was inhabited by the so-called *Tumulus culture* coming from the west (these people built fortifications) (Gyulai 1996a).

Early (Hallstatt) (900-500 BCE), Middle (Skythian) (700-500 BCE) and Late Iron Age (La Tène) (500-100. CE)

During the Iron Age (9th CENT. BCE – 1st CENT. CE), the climate in the Carpathian Basin changed to continental. This period was associated with the spread of *Celtic* populations from the south-east with relatively few plant remains (Gyulai 1996b). However, common wheat and spelt were the most important cereals, with some reduced remains of einkorn, barly, millet, flax, lentils, peas, poppy, cucumbers, and fruits including plums, peaches and grapes.

Roman Age (1st – 5th Cent. CE) and Barbaricum

The Roman Age in *Pannonia* brought new crops, farming implements and technologies. Walnuts, plums, apricots, peaches and grapes were cultivated by the Romans (Figures 1, 2) and they did not disappear from Pannonia with the fall of the Roman Empire and continue to the present time (Gyulai *et al.* 2009). Barbars, like *Sarmatas,* inhabited the Great Hungarian Plain in this time east to Roman province Pannonia. Hulled wheats (einkorn and emmer) and barley were characteristic of the culture along with naked common wheat and dwarf wheat, rye, millet, as well as peas, lentils and faba beans (Figures 1, 2).

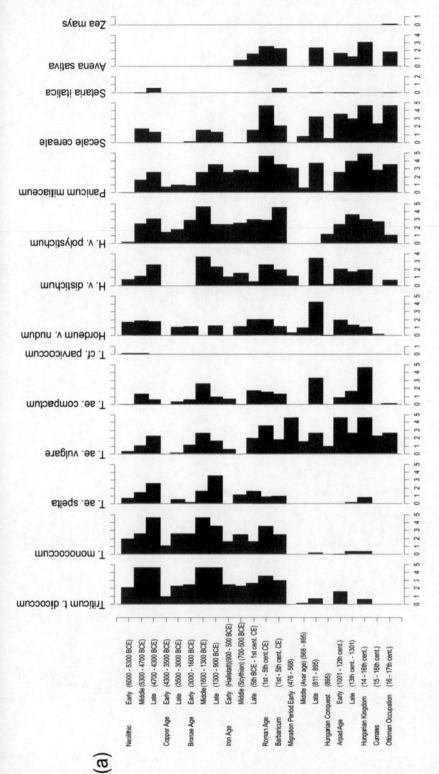

Figure 2 (Continued).

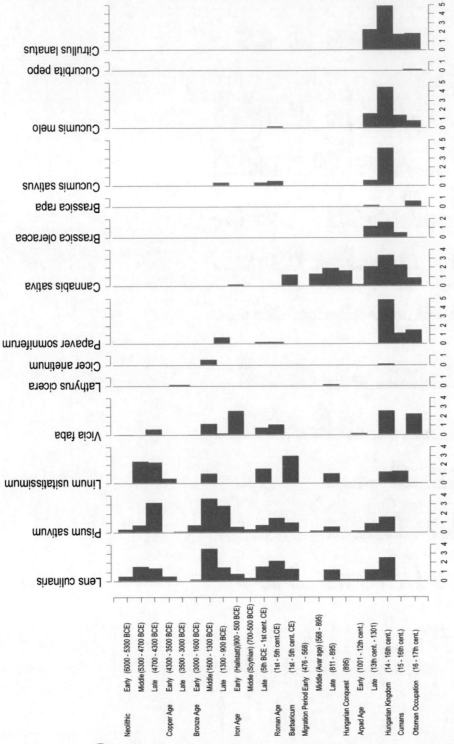

(b)

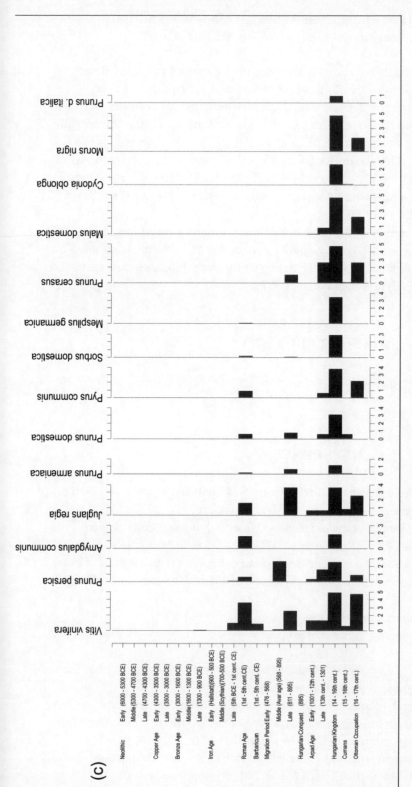

Figure 2. Distribution frequencies (Juggins 2007) of plant remains (log 1-5) in Carpathian Basin (Hungary) from the Neolithic (8000 BP) to the late Medieval Age (17th CENT. CE), (a) cereals, (b) garden plants and (c) fruits.

Early (476-568), Middle (Pre-Hungarian Avarage) (568-895) and Late (Caroling Age) Migration Period (811-895) to The Middle Ages

During the Migration Period, tribes and nations passed through the Carpathian Basin (the Crossroads of Europe) and destroyed the Roman civilization and agriculture, however, grape and fruit production were adapted by the new nations as shown by the excavations near Lake Balaton (Figures 1, 2). Climatic factors may also have stimulated this mass-movement. Huns and Avars cultivated millet and barley (Györffy 1977). No oats were found in archaeogenetical sites until the Hungarian Conquest (895). The cultivation of common wheat and rye became intensive from the 12th–13th centuries (Figures 1, 2). The 14th–15th centuries opened a new period in agriculture with new plants from the Mediterranean (Facsar 1970; Hartyányi and Nováki 1975; Füzes 1990).

The prosperity of cereal production ended under the Ottoman occupation in the 16th century. Later, new plants like corn (*Zea mays*) appeared in the 17th century from the New World and the grain and fruit production in the Carpathisan Basin become intensive again and was marketed in all of Europe (Gyulai and Gyulai 2009).

ACKNOWLEDGMENTS

The authors wish to thank Dr. Luther Waters Jr. and Dr. Gábor Gyulai for the comments on the manuscript.

REFERENCES

Beijerinck W (1947) Zadenatlas der nederlendische Flora. H. Veenman, Wageningen, 316 p.

Brecher Gy (1960) *Seed morpology* (in Hungarian). Mgk, Budapest.

Facsar G (1970) Vergleichende morfologische Untersuchungen der Samen von Gartenrebensorten I.. *Botanikai Közlememények* 57/3.

Füzes M (1990) Plant remains from the early phase of plant cultivation (Neolithic and Copper Age) Hungary. An archaeogenetical outline. *Tapolcai Városi Múzeum Közleményei* 1: 139-238.

Györffy I (1977) *King St. Stephanus and his Kingdom.* Gondolat, Budapest.

Gyulai F (1993) Environment and Agriculture in Bronze Age Hungary. *Archaeolingua,* Budapest, 59 p.

Gyulai F (1996a) Plant findings and food remnants of the Late Bronze Age settlement at Balatonmagyaród-Hídvégpuszta. *Zalai Múzeumok* 6: 169-195.

Gyulai F (1996b) Umwelt und Pflanzenbau in Transdanubien während der Urnenfelder-, Hallstatt- und Latènekultur. In: Jerem E, Lippert A (eds) *Akten des Internationalen Symposiums,* Sopron, 10-14. Budapest, *Archaeolingua*: 127-136.

Gyulai F (2000) The Changing of the Cereals-Diversity in the Carpathian Basin. Archaeobotanical Review.) In: Gyulai F (ed) *Symp. Jánossy Andor*: 278–287.

Gyulai F (2007) Seed and fruit remains associated with neolithic origins in the Carpathian Basin. In: Colledge S., Conolly J (eds) The Origins and Spread of Domestic Plants in

Southwest Asia and Europe. University College London Institute od Archaeology Publications. Left Coast Press, Walnut Creek, CA, 446 p.,125-140.

Gyulai F and Gyulai G (2009) Medieval viticulture in Hungary. In: Dénesi T, Csoma Zs (eds) „ad vinum disert..." Budapest, 2009, 256 p., 27-40.

Gyulai G, Tóth Z, Szabó Z, Gyulai F, Lágler R, Kocsis L, Heszky L (2009) Domestication Events of Grape (*Vitis vinifera)* from Antiquity and the Middle Ages in Hungary from growers' viewpoint. *Hung Agric Res* 2009/3-4: 8-12.

Hartyányi PB and Nováki Gy (1975) Samen- und Fruchtfunde in Ungarn von der Jungsteinzeit bis zum 18. Jahrhundert. *Agrártörténeti Szemle* 17: 1–22. Supplementum.

Hartyányi PB, Nováki Gy, Patay Á (1967–68) Samen- und Fruchtfunde in Ungarn von der Jungsteinzeit bis zum XVIII. Jahrhundert I. *Magyar Mezőgazdasági Múzeum Közleményei*: 5–85.

Hartyányi PB, Máthé BSz (1980) An die Oberfläche 1976 gebrachte pflanzliche Überreste einer Wohnsiedlung in Szilhalom bei Berettyóújfalu aus der späteren Zeit des Neolithikums. *Magyar Mezőgazdasági Múzeum Közleményei 1980*: 125–145.

Juggins S (2007) C2 Version 1.5. *User guide*. Software for ecological and palaeoecological data analysis and visualisation. Newcastle University, Newcastle upon Tyne, UK.

Rovner I and Gyulai G (2007) Computer-Assisted Morphometry: A New Method for Assessing and Distinguishing Morphological Variation in Wild and Domestic Seed Populations. *Journal of Economic Botany* 61/2: 1–19.

Schermann Sz (1966) Magismeret I-II. [Seed identification I-II.]. Akadémiai Kiadó, Budapest, 861+208 p.

Schoch W, Pawlik B, Schweingruber FH (1988) Botanische Makroreste. Ein Atlas zur Bestimmung häufig gefundener und ökologisch wichtiger Pflanzensamen. Paul Haupt, Bern, 227 p.

Zohary D and Hopf M (2000) Domestication of Plants in the Old World. Oxford University Press, New York, 316 p.

In: Plant Archaeogenetics
Editor: Gábor Gyulai, pp. 41-48

ISBN: 978-1-61122-644-7
© 2011 Nova Science Publishers, Inc.

Chapter 5

MORPHOGENETICS OF ANCIENT *Vitis* SEEDS (3RD AND 11TH - 15TH CENTS, HUNGARY) - A GENOTYPE RECONSTRUCTION

Gábor Gyulai[1,], Renée P. Malone[2], Luther Waters Jr.,[3]
László Heszky[1] and Erzsébet Kiss[1]*

[1]St. Stephanus University, Institute of Genetics and Biotechnology,
Gödöllő, 2103, Hungary
[2]Dublin Institute of Technology,
School of Food Science and Environmental Health, Dublin 1, Ireland
[3]Department of Horticulture, College of Agriculture,
Auburn University, Alabama 36849, USA

ABSTRACT

Morphogenetics of grape (*Vitis vinifera*) seed remains from Roman (3rd CENT CE) (*current ere*) and Medieval (11th to 15th CENT CE) archeological sites in Hungary are discussed in this Chapter. Seeds were analyzed by LM (Light Microscopy) and SEM (*Scanning Electron Microscopy*). Excavation sites included Budapest (*Aquincum*; 3rd CENT, Hungary) and Keszthely (Fenékpuszta; 5th CENT, Hungary) of Roman Age; and Győr (Ece, 11th-12th CENT, Hungary), Debrecen (13th CENT, Hungary) and the King's Palace of Árpád Dynasty at the Castle of Buda, Budapest (15th CENT, Hungary) of the Middle Ages. Ancient seeds were compared to thirty current grape varieties of similar seed size, shape and morphology. The current blue grape variety *Vitis vinifera* cv. '*kék bakator*' (*syn.:'Blue Bocca d'Oro*'; '*aranybogyó*', '*golden grapes*') (# 25, Figure 4) was found most similar in seed morphology to one of the ancient samples (15th CENT Debrecen, Hungary) which indicates the antiquity of this cultivar.

---

* E-mail: gyulai.gabor@mkk.szie.hu

INTRODUCTION

Species of the plant family *Vitaceae* are woody climbers comprising of 13 – 17 genera (*Acareosperma, Ampelocissus, Ampelopsis, Cayratia, Cissus, Clematicissus, Cyphostemma, Leea, Muscadinia, Nothocissus, Parthenocissus, Pterisanthes, Pterocissus, Rhoicissus, Tetrastigma, Yua* and *Vitis*). The genera include about 700 species (Arnold *et al.* 2002; Chen and Manchester 2007).

Figure 1. Excavation sites from the antiquity at the Roman Camp of *Aquincum* (Budapest, Hungary) 3rd CENT CE (from the Photo Archives of the Aquincum Museum, Budapest, Hungary).

The genus *Vitis* consists of about 60 inter-fertile species, including about fifteen species of agronomic importance (Terral *et al.* 2010). Of them, *Vitis vinifera* ($2n = 4x = 38$) is the only species which is indigenous to Eurasia, with a relatively small nuclear (nDNA) genome size ($0.475 - 0.5 \times 10^9$ DNA base pair), and a regular size (160,928 bp, NCBI #DQ424856) of chloroplast cpDNA (Jansen *et al.* 2006) and mtDNA (773,279 bp; NCBI #FM179380). Most genera of family *Vitaceae* have $2n = 38$ chromosomes ($n = 19$), however species of *Muscadinia, Ampelocissus, Parthenocissus*, and *Ampelopsis* have $2n = 40$ chromosomes ($n = 20$), and species of the genus *Cissus* has $2n = 24$ chromosomes ($n = 12$).

The wild, dioecious ancestor form of *V. silvestris* still co-exists in Eurasia and North Africa with the cultivated *V. vinifera* which has hermaphrodite flowers (Arnold *et al.* 2005; This *et al.* 2006). Today, thousands of cultivars have been developed, which are generally classified in main groups according to their final production; as wine grapes, table grapes, including current seedless grapes, and raisins (Terral *et al.* 2010).

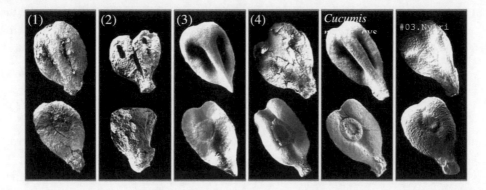

Figure 2. Scanning Electron Micrographs of ancient grape (*Vitis vinifera*) seeds excavated in Hungary. *(1)* 3rd CENT seed from Roman Villa at Aquincum (Budapest, Hungary). *(2)* 5th CENT seed from Roman Age (Keszthely - Fenékpuszta, Hungary). *(3)* 11st-12th CENT seed form a vineyard near Győr (Ece, Hungary). *(4)* 13th CENT seed from Debrecen (Hungary). *(5)* 15th CENT seed from the King's Palace of Árpád Dynasty in the Castle of Buda (Budapest, Hungary). *(6)* Seed of current grape cv. *'kék bakator'*. Ventral (up) and dorsal (down) view (20X).

Genetically, dioecy in wild grapes are encoded by a single gene; female individuals are homogametic, carrying homozygous recessive pistil-suppressor alleles (Su^mSu^m) which suppress the development of anthers (and pollen). Male plants are heterozygous (Su^FSu^m), carrying a dominant pistil-supressing Su^F allele. The shift, under domestication, to a bisexual (hermaphrodism) flower was attained by a single mutation to Su^+, which is also dominant over Su^m, resulting in two genotypes of hermaphroditc grape types (Su^+Su^m and Su^+Su^+) (Zohary and Hopf 2000; McGovern 2004).

Figure 3. The oldest, 8000-yr-old *Vitis* seeds excavated in Hungary (Berettyóújfalu, *Kőrös Culture*, 6000 BCE) (F Gyulai *et al*. this volume). Wild grapes (*Vitis sylvestris*) (seed sizes: 4.5 - 5.0 mm; Photo by Zsigmond G Gyulai) (notice the ventral *infolds*, the dorsal *chalaza;* and short seed *beak* characteristic to wild grapes).

European grape growing in North America formed several hybrids with native *Vitis* species. Some of these hybrids became resistant to *Phyloxera* (an insect pest), which devastated European vineyards in the 1880s, and supplied resistant rootstocks for replantations. This event indicates that the grape genome diversity has been narrowed twice; first by the Biblical flood, followed by the replantation of Noah 'the first vintner' (Genesis 9) on Mount Ararat (McGovern 2004), and second by *Phyloxera* (This *et al.* 2006). Unlike the genome of dioecious *V. sylvestris*, genetic diversity of grapes has been narrowing continuously as the result of vegetative propagation either by rooting of twigs, or by grafting.

MATERIALS AND METHODS

Excavated and wet-sieved sediment samples were processed by floatation followed by seed sorting and identification in the laboratory according to Schermann (1966).

For SEM analysis, seeds were air dried, fixed in glutaraldehyde (5% w/v in phosphate buffer 0.07 M, pH 7.2) and washed three times in the same buffer for 10 minutes. They were then desiccated in acetone concentration series (10-50-70-90-100 %), dehydrated at the CO_2 critical point (Blazers CDC 020), and covered with gold (30 nm). Samples were examined and photographed using a TESLA BS-300 scanning electron microscope as described by Gyulai *et al.* (1992) and Szabó *et al.* (2007).

For LM analysis (Greguss 1967; Schermann 1966), a Leica microscope (# 301-371.010) was used. Seeds of thirty current *Vitis* cultivars (Figure 4) currently grown in Hungary were used for comparative analyses (Chen and Manchester 2007; Hardie *et al.* 1996; Mangafa and Kotsakis 1996; Facsar 1970; Terpó 1976).

Ancient DNA (aDNA) were also extracted from the seeds according to Gyulai *et al.* (2001, 2006), Bisztray *et al.* (2004), and amplified by WGA (Genomplex, Whole Genome Amplification, Sigma WGA-2) with a 5 - 9 fold amplification rate of total genomes, and analyzed by *Vitis*-specific primer pairs (Manen *et al.* 2003; Szabó *et al.* 2007).

Table 1. The oldest seed remains of *V. vinifera* excavated in Hungary
(F Gyulai *et al.* this volume)

Ages	Excavation sites (Hungary)	# seeds
1300 BCE	Sopron	some
3rd CENT (Roman age)	Aquincum (Budapest, Kaszásdűlő)	6108
5th CENT (Roman age)	Keszthely-Fenékpuszta	33
11-12th CENT	Győr (Ece)	28
13th CENT	Debrecen (Kölcsey Cultural Centre)	24
15th CENT	Budapest (King Palace)	210,686

CONCLUSION

The oldest (8400 BP) wild grape (*Vitis sylvestris*) seeds (about 3 mm long) were excavated in Turkey, at Nevali Çori (NÇ) located near the Turkish city of Urfa (37°60'N, 38°70'E, 490m above sea level) on the slope of a Euphrates side valley, Hilvan province (Hauptmann 1997; Pasternak 2008). The oldest wild grape seeds excavated in Hungary (Berettyóújfalu, *Kőrös Culture*, 6000 BCE) were 8000-yr-old (*Vitis silvestris*) (Figure 3).

The first convincing evidence of *Vitis vinifera* seeds with indications of grape cultivation were also uncovered in Turkey at Kurban Höyük (5700 – 5200 BP non-calibrated radiocarbon time) (Zohary and Hopf 2000; McGovern 2004), and the early Bronze Age (3200 – 1900 BP) sites along the Jordan Valley, at *Tell Shuna* (Jordan; Chalcoitic), *Jericho* (Cisjordan; early Bronze Age), and *Arad* (Israel, early Bronze Age) (Jacquat and Martinoli 1999). Ancient grape seeds were also excavated at Semma (Sudan) 3500 BP (Zeist 1983).

The earliest evidence of wine production (jars from Godin) was found in Iran (Hajji Firuz Tepe site in the Zagros Mountains) about 7400-7000 BP (This *et al.* 2006) and 5500 - 4900 BP (McGovern 2004). Later, Greek, Latin, and Egypt vine amphoras with gelified vine remains were also found in the hulls of sunken ships (McGovern 2004), similar to the famous shipwreck remains at Uluburun, near Kas (Turkey) (Ward 2003). Grape cultivation gradually spread to Mesopotamia, Assyria, and Egypt (about 5500 – 5000 BP), and further west along the Mediterranean to Phoenicia, Greece, North Africa and then to the entire Roman Empire north to Pannonia (Hungary) and German tribes. Viticulture also spread eastward along the Silk Road and it reached China and Japan in 3200 BP (Rivera and Walker 1989).

The grape was introduced to the Americas by European colonists starting from the 16th CENT after either the early Chinese explorer Zheng He (1405 – 1435), or the Columbus voyages (first: Aug. 3 1492 to March 15 1493; second: Sept. 25 1493 to June 11 1495; third: May 30 1498 to Nov. 15 1500; fourth: May 11 1502 to Nov. 7 1504).

The first plantations in North America were established on the West Coast by Spanish missionaries and later by viticulturists like Ágoston Haraszty, who is considered the father of California's grape-growing industry. Haraszty imported 200,000 grape cuttings from Europe after 1849, including grape varieties from his native Hungary, and developed over half a million California acres to viticulture, making wine growing second only to orange production in the state's agricultural economy. In recognition of his merits, Haraszty was named California's State Commissioner of Viticulture (Sisa 2006).

In Hungary, the earliest 8000-yr-old wild grape (*Vitis sylvestris*) seeds were found at Berettyóújfalu (Figure 3), and Tiszapolgár (5300 BCE). The earliest *Vitis vinifera* seeds are from Sopron (Hungary) (1300 BCE), which dates the origins of grape cultivations to the late Bronze Age (Table 1).

The earliest wine residue dates back to 700 BCE (Fehérvárcsurgó, Hungary), which places the beginnings of wine-making to the Iron Age. Thus, there is evidence that both grape cultivation and wine making date to well before to the Roman period in Hungary.

To conclude, seeds of *Vitaceae* were easily identified from a suite of unique and distinctive morphological characters (particularly a pair of ventral infolds and a dorsal chalazal scar) (Chen and Manchester 2007). Based on seed morphogenetics, the 15th CENT grape seeds (Figure 2, *#5*) were similar to the currently grown grape variety '*kék bakator*' ('Blue Bocca d'Oro') (Figure 2. *#6;* Figure 4. *#25*), which is one of the oldest variety grown in Hungary, as the etymology of its name *Bocca d'Oro* (*aranybogyó*) suggests (Gyulai *et al.* 2009; Güner *et al.* 2009). Other seed samples, with obvious ancient type of short seed beaks from the Roman (3rd, and 5th CENTS, *#1* and *#2* Figure 2) and medieval age (13th CENTs, *#4* Figure 2) showed no such similarity to any of the thirty currently grown grape varieties analyzed (Figure 4). Seed sample from the 11th – 12nd CENT (*#3* Figure 2) showed incomparably unique grape seed types.

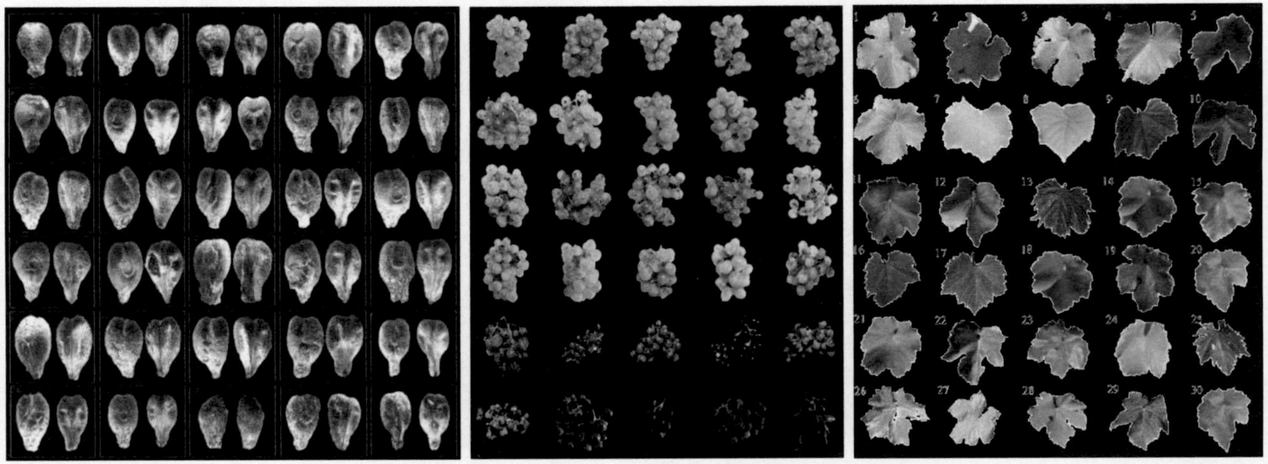

(a)

(b)

(c)

Figure 4. Dorsal and ventral view of *(a)* seed, *(b)* grape and *(c)* leaf morphology of current grapes *(Vitis vinifera)* grown in Hungary used for morphological reconstructions of ancient grapes. *cvs: (1) Rajnai rizling NI – 378. (2) Leányka. (3) Zöldszilváni. (4) Ezerfürtü. (5) Juhfark. (6) Chasselas blanc K-15. (7) Kunleány (V. amurensis x V. vinifera, Tamássy & Koleda 1960). (8) V. riparia x V. rupestris. (9) Narancsizü. (10) Fehér lisztes. (11) Mirkpvacsa. (12) Hárslevelü. (13) Változó góhér. (14) Sárfehér. (15) Kéknyelü. (16) Csabagyöngye. (17) Mátyás király. (18) Fehér járdovány. (19) Kossuth. (20) Piros góhér. (21) V. vinivera 'Aramon' x V. riparia 143 B. (22) Bakó. (23) Visnivi rami. (24) Kékfrankos. (25) Kék bakator. (26) Oportó. (27) Szürkebarát D 34. (28) Kismis vatkana. (29) Piros szlanka. (30) Suvenir* (five varieties in each line) (Photo by R Lágler and L Kocsis at the same time Sept. 17. 2006, Keszthely, Hungary).

REFERENCES

Arnold C, Rosetto M, McNalley J, Henry RJ (2002) The application of SSRs characterized for grape (*Vitis vinifera*) to conservation studies in *Vitaceae*. *Amer J Bot* 89: 22-28.

Arnold C, Schnitzler A, Douard A, Peter R, Gillet F (2005) Is there a future for a wild grapevine (*Vitis vinifera* subsp. *silvestris*) in the Rhine Valley? *Biodiversity and Conservation* 14: 1507-1523.

Bisztray GyD, R Bacsó, P Bodor, G Facsar, F Gyulai, I Velich (2004) Archaeobotanical and genetical methods to analyse 600-years-old seeds of horticultural plants. Book of Abstracts 5th IVCHB Symposium, *In Vitro Culture and Horticultural Breeding*, 12-17. September 2004, Debrecen, Hungary, p. 212.

Chen I, Manchester SR (2007) Seed morphology of modern and fossil *Ampelocissus* (*Vitaceae*) and implications for phytogeography. *Amer J Bot* 94: 1534-1553.

Facsar G (1970) Habitus studies on seeds *Vitis vinifera* L. sorts. *Acta Agr Hung* 19: 403-406.

Greguss P (1967) Fossil gymnosperm woods in Hungary from the Permian to the Pliocenes. Akadémiai Kiadó, Budapest.

Güner A, Gyulai G, Tóth Z, Başli GA, Szabó Z, Gyulai F, Bittsánszky A, Waters L, Heszky L (2009) Grape (*Vitis vinifera)* seeds from Antiquity and the Middle Ages Excavated in Hungary - LM and SEM analysis. *Anadolu Univ J Sci and Technol* 10: 205-213.

Gyulai G, Janovszky J, Kiss E, Lelik L, Csillag A, Heszky LE (1992) Callus initiation and plant regeneration from inflorescence primordia of the intergeneric hybrid *Agropyron repens* (L.) Beauv. x *Bromus inermis* Leyss. cv. *nanus* on a modified nutritive medium. *Plant Cell Rep* 11: 266-269.

Gyulai G, Magda A, Kiss J, Gyulai F, Holly L, Heszky L (2001) DNS izolálás és PCR-amplifikáció 700 éves növény magvakból. VII. Növénynemesítési Tudományos Napok, Budapest, p.89.

Gyulai G, Humphreys M, Lágler R, Szabó Z, Tóth Z, Bittsánszky A, Gyulai F, Heszky L (2006) Seed remains of common millet from the 4th (Mongolia) and 15th (Hungary) centuries; AFLP, SSR, and mtDNA sequence recoveries. *Seed Science Research* 16: 179-191.

Gyulai G, Tóth Z, Szabó Z, Gyulai F, Lágler R, Kocsis L, Heszky L (2009) Domestication Events of Grape (*Vitis vinifera)* from Antiquity and the Middle Ages in Hungary from growers' viewpoint. *Hung Agric Res* 2009/3-4: 8-12.

Hardie WJ, O'Brien TP, Jaudzems VG (1996) Morphology, anatomy and development of the pericarp after anthesis in grape, *Vitis vinifera* L. *Australian J of Grape and Wine Research* 2(2): 97-142.

Hauptmann H (1997) Nevali Çori. *The Oxford Encyclopedia of Archaeology in the Near East* 4: 131-134.

Jacquat C, Martinoli D (1999) *Vitis vinifera* L.: Wild or cultivated? Study of the grape pips found at Petra, Jordan; 150 B.C. – CE 40. *Veget Hist Archaeobot* 8: 25-30.

Jansen RK, Kaittanis C, Saski C, Lee S-B, Tomkins J, Alverson AJ, Daniell H (2006) Phylogenetic analyses of *Vitis* (*Vitaceae*) based on complete chloroplast genome sequences: effects of taxon sampling and phylogenetic methods on resolving relationships among rosids. *BMC Evolutionary Biology* 6:32 doi:10.1186/1471-2148-6-32.

Manen J-F, Bouby L, Dalnoki O, Marinval M, Turgay M, Schlumbaum (2003) Microsatellites from archaeological *Vitis vinifera* seeds allow a tentative assignment of the geographical origin of ancient cultivars. *J Archaeological Science* 30: 721-729.

Mangafa M, Kotsakis K (1996) A new method for the identification of wild and cultivated charred grape seeds. *J Arch Sci* 23: 409-418.

McGovern PE (2004) Ancient wine: the search for the origins of viniculture. Princeton University Press.

Pasternak R (2008) Investigations of Botanical Remains from Nevali Çori PPNB, Turkey: A Short Interim Report. http://www.bioversityinternational.org.

Rivera ND, Walker MJ (1989) A review of palaeobotanical findings of early *Vitis* in the Mediterranean and the origins of cultivated grapevines, with special reference to new pointers to prehistoric exploitation in the western Mediterranean. *Rev Palaeobot Palynol* 61: 205-237.

Schermann Sz (1966) Seed morphology (Magismeret, in Hungarian), Vol. I (pp. 861) and II (pp. 209), Akadémiai Kiadó, Budapest.

Sisa S (2006) The Spirit of Hungary, Vista Books, Morristown New Jersey, USA. ISBN 0-9628422-0-6.

Szabó Z, Gyulai G, Kovács L, Tóth Z, Lágler R, Bittsánszky A, Kocsis L (2007) Ancient DNA analysis and morphology of grape seeds from antiquity and the middle ages excavated in Hungary. *Botany and Plant biology Joint Congress*, Chicago, Illinois, USA, July 7-11. ID: 2119.

Terpó A (1976) The carpological examination of wild-growing vine species in Hungary. *Acta Bot Hung* 22: 209-247.

Terral J-F, Tabard E, Bouby L, Ivorra S, Pastor T, Figueiral I, Picq S, Chevance J-B, Jung C, Fabre L, Tardy C, Compan M, Bacilieri R, Lacombe T, This P (2010) Evolution and history of grapevine (*Vitis vinifera*) under domestication: new morphometric perspectives to understand seed domestication syndrome and reveal origins of ancient European cultivars. *Annals of Botany* 105: 443-455.

This P, T Lacombe, MR Thomas (2006) Historical origins and genetic diversity of wine grapes. *Trends in Genetics*, 22: 511-519.

Zeist van W (1983) Fruits of foundation deposits of tow temples. *J Archaeol Sci* 10: 351-354.

Zohary D and Hopf M (2000) *Domestication of plants in the Old World*. Oxford: Oxford University Press. ISBN 0-19-850356-3.

Ward C (2003) Pomegranates in eastern Mediterranean contexts during the Late Bronze Age. *World Archaeol* 34: 529-541.

II. aDNA Extraction and Molecular Probing of Archaeological Specimens

In: Plant Archaeogenetics
Editor: Gábor Gyulai, pp. 51-68

Chapter 6

DNA EXTRACTION FROM SEED REMAINS OF COMMON MILLET (*Panicum miliaceum*) (4TH AND 15TH CENTS)

Gábor Gyulai[1,], Mervyn O. Humphreys[2] and Richárd Lágler[1]*
[1]St. Stephanus University, Institute of Genetics and Biotechnology,
Gödöllő 2103, Hungary
[2]Institute of Biological, Environmental and Rural Sciences,
Aberystwyth University, Wales, UK

ABSTRACT

Ancient DNA (aDNA) was extracted from seed remains of common millet (*Panicum miliaceum*) excavated from sites of 4th CENT CE (Darhan, Mongolia) and 15th CENT CE (*current era*) (Budapest, Hungary). aDNA samples were analyzed by AFLP, SSR, ISSR, mtDNA and CAP-mtDNA methods.

Agarose gel electrophoresis revealed that extensive DNA degradation had occurred in the 4th CENT millet resulting in only 2 AFLP fragments (1.2 %), compared to the 15th CENT millet with 158 fragments (40.0 %) and current millet cv *Topáz* with 264 fragments (100 %). Eight AFLP fragments were sequenced after reamplification and cloned in the medieval millet.

Four microsatellites (SSR) were amplified in the medieval millet at the *gln*4, *sh*1, *rps*28 and *rps*15 gene loci and compared to current variety cv *Topáz*. No SSR was amplified in the 4th CENT sample. After sequencing, the medieval SSR fragments showed identical alignments with current millet sequences with only one SNP (*single nucleotide polymorphism*) at the 6th position (A to G) of the *rps*28 locus.

The mtDNA specific primer pair amplified an 1117 bp fragment at the 5S-18S rDNA locus in the 15th CENT millet and current millet cv *'Omskoje'*. In the CAP-analysis of this fragment, there were no observed changes in the DNA sequences at restriction sites of *Taq*I, *Bsu*RI, *Hinf*I, *Mbo*I, *Alu*I and *Rsa*I giving the same fragment pattern on agarose gel

* E-mail: gyulai.gabor@mkk.szie.hu

with the same numbers and lengths of digested fragments. After sequencing of *Mbo*I fragments of the 5S-18S rDNA, no SNP was observed in the nucleotide sequences. Authenticity of aDNA sequences were proved by BLAST analysis in each case.

In the ISSR analysis, seven ISSR primers and combinations amplified 15 alleles in the 15th CENT millet and twenty current millet cultivars used for comparative analysis. Phenotype reconstruction was carried out based on ISSR dendrogram, which revealed that medieval millet might had been very similar to current variety *'Omskoje'*.

Keywords: *A*ncient DNA (aDNA), CAP (*c*leaved *a*mplified *p*olymorphic DNA), fAFLP (*f*luorescent *a*mplified DNA *f*ragment *l*ength *p*olymorphism), genotype, ISSR (*i*nter *s*imple *s*equence *r*epeat), microsatellites (SSR – *s*imple *s*equence *r*epeat), mitochondrial DNA (mtDNA at 5S-18S rDNA locus), phenotype, SNP (*s*ingle *n*ucleotide *p*olymorphism).

INTRODUCTION

Common millet *(P. miliaceum*, 2n = 4x = 36) is one of the most ancient grain crops of humanity with the oldest historical reports from 5000 - 3200 BCE (*Before Current Ere*) (Ho 1977). However, the oldest plant remains with proven human activity have revealed cereal seeds as wild barley (*H. spontaneum*) and wild emmer (*Triticum dicoccoides*) from 23,000 - 19,000 BP at Ohalo II., river Jordan (Nadel *et al.* 2006; Weiss *et al.* this volume). The 17,310 ± 310 BP site in Korea (Chungbuk National University, South Korea) revealed the first ancient rice (*Oryza sativa*) seed remains with extractable and amplifiable amount of aDNA (Suh *et al.* 2000). The *Hoabinhian* culture also produced various plant remains from approximately 12,000-8000 BP (Gorman 1969; Walters 1989). *Panicum* became typical foods of the Sumer and Northern India together with barley (*H. vulgare*) in about 2500 BCE. For the nations of steppic Scythia, such as the Hungarians or Celtics in 2000 BCE, the common millet was the first crop that produced two harvests in one year. In the ancient Chinese 'Book of Poetry' (*Shih Ching*), written about 1000 - 500 BCE, nine poems mention common millet (Keng 1974). Common millet spread from the Steppes through Europe via tribes of the Celts, Huns, Avars and Hungarians, and also through the region of the 'Fertile Crescent' and Africa (Harlan 1971; Vaughan *et al.* 2007). It became the *milium* of Romans (Smith 1977). Millet was introduced to North America in the 17th century (Colosi and Schaal 1997), and new cultivars were registered recently (Baltensperger *et al.* 1997).

In this Chapter, we present the analysis of aDNA of 1,600-year-old common millet seed remains from Mongolia (4th CENT), and a 600-year-old seeds from a 15th CENT site of Hungary together with current cultivars with a final aim of tracking molecular domestication and phenotype reconstruction of ancient millets.

MATERIALS AND METHODS

Seed Samples

Seed remains of common millet (*P. miliaceum*) from the 4th CENT site (3rd grave in Darhan, Mongolia, 1969) (Tseveendorj and Sugar 1994); and the 15th CENT site (8th well, Mansion Teleki, King's Palace of Árpád Dynasty, Budapest, Hungary) (Gyulai *et al.* 2006) were analyzed. Wet-sieved sediment samples were processed by floatation followed by seed sorting and identification in the laboratory.

For comparative analysis, twenty current common millet cultivars (ABI, Tapiószele, Hungary) were included (Figure 7): *(1)* Tápiószele-D; RCAT073416. *(2)* Tápiószele-C; RCAT073585. *(3)* Tápiói barna; RCAT017521. *(4)* Tápiószentmártoni (lr); RCAT017513. *(5)* Tápiószele-B; 00185/01. *(6)* Debreceni barna; RCAT017280. *(7)* Fertődi 10.D; RCAT017272. *(8)* Püski (lr); RCAT017296. *(9)* Rábaszentandrási (lr); RCAT017297. *(10)* Bolgár-159; RCAT017267. *(11)* Fertődi piros; RCAT017291. *(12)* Kecskeméti (lr); RCAT017527. *(13)* Omskoje; 02546/00. *(14)* Jászberényi (lr); RCAT017555. *(15)* Császárréti; RCAT017277. *(16)* Nyiregyházi (lr); RCAT017526. *(17)* Tápiószele-A; RCAT017509. *(18)* Fertődi fehér; RCAT017290. *(19)* Martonvásári-3; RCAT017285. *(20)* Mesterházi (lr); RCAT017494.

Aseptic Culture

Seed remains were washed with detergent (3 min) and rinsed three times with distilled water (3 min), followed by surface sterilization with ethanol (70 %) for 1 min and a commercial bleaching agent (8% NaOCl) for 1 min; followed by three rinses with sterile distilled water according to general tissue culture technique, and incubated for a month in aseptic tissue culture media F6 (Gyulai *et al.* 1992, 2003) to eliminate cross contaminations before DNA extraction. Exogenously and endogenously contaminated seeds infected by fungi and bacteria were eliminated from further study.

DNA Extraction

Individual seeds were ground in an aseptic mortar with liquid nitrogen in a laminar air flow cabinet of an archaeogenetic lab of the St Stephanus University (Gödöllő, Hungary), followed by aDNA extraction using the CTAB (cethyltrimethylammonium bromide) method according to Szabó *et al.* (2005) and Lágler *et al.* (2005) based on Murray and Thompson (1980); and Doyle and Doyle (1990).

DNA samples were treated by RNase-A (from bovine pancreas) (Sigma, R-4875), for 30 min at 37° C. Seed-DNA of current cultivars was also extracted in each case.

The quality and quantity of extracted DNA were measured (2 µl) by a NanoDrop ND-1000 UV-Vis spectrophotometer (NanoDrop Technologies, Delaware, USA – BioScience, Budapest, Hungary).

DNA samples were adjusted (except AFLP) to a concentration of 30 ng/µl with ddH$_2$O and subjected to PCR amplification (Gyulai *et al.* 2001; Bisztray *et al.* 2004).

PCR Amplification

A minimum of two independent DNA preparations of each sample were used for PCR amplification. A negative control which contained all the necessary PCR components except template DNA was included in PCR runs. Pooled DNA extracted from the non-infected seeds (0.1 g) was used.

PCR Primers and SSR Analysis

The following primer pairs (Chin *et al.* 1996) were applied for the amplification of SSR loci of *gln*4 gene (cytoplasm *glutamine synthetase4*) (*agc aga acg gca agg gct act* and *ttt ggc aca cca cga cga*) (D14577), including SSR (ttgcg)$_2$; *sh*1 gene (endosperm-specific *sucrose synthase: shrunken*1) (*atc gaa atg cag gcg atg gtt ctc* and *atc gag atg ttc tac gcc ctg aag t*) (AF544115) including SSR (aag)$_6$; *rps*28 gene (structural *ribosomal protein S*28) (*aga cga acc cac cat cat ctt tc* and *cgc ttg gca tct cca tgt ata tct*) (AW424565) including SSR (tc)$_8$ and *rps*15 (AV062092) (*aag aag aaa gag aag aag cac ggg* and *gga cag ctc gta gta taa cct gcg*) including SSR (cag)$_5$.

AFLP Analysis

Undiluted aDNA samples (5.5 µl) were subjected to fAFLP (*f*luorescent *a*mplified DNA *f*ragment *l*ength *p*olymorphism) analysis using CGE (*C*apillary *G*el *E*lectrophoresis) by ABI PRISM 3100 Genetic Analyzer (Applied Biosystems) following the method of Vos *et al.* (1995) with modifications according to Hodkinson *et al.* (2000), Cresswell *et al.* (2001), Skøt *et al.* (2002), and Gyulai *et al.* (2005, 2006). For digestion-ligation reactions, pairs of *Eco*RI - *Mse*I restriction endonucleases (REases) were used. The sequences of the preselective primer pairs were: Eco-A: *gac tgc gta cca att c-a*, and Mse-C: *gat gag tcc tga gta a-c*. For selective amplification 24 primer combinations were used with JOE (green) fluorescent labeled *Eco-primers. In primer combinations 1 to 12 the primer Mse-CAC was combined with labeled primers of *Eco -aaa, -aac, -aag, -aat, -aca, -acc, -agg, -act, -aga, -agc, -agg, -agt. In primer combinations 13 to 24 the labeled primer *Eco-AGT was combined with primers of Mse -caa, -cag, -cat, -cca, -ccc, -ccg, -cct, -cga, -cgc, -cgg, -cgt, -cta according to Gyulai *et al.* (2006). All oligonucleotides were supplied by Sigma Genosys, and enzymes by Roche Diagnostics. PCR-amplified AFLP fragments were subsequently denatured at 98 °C for 5 min, and directly forwarded to an ABI PRISM 3100 Genetic Analyzer (Applied Biosystems) using a G5 filter set in two repetitions, and analyzed by ABI PRISM Genotyper 3.7 NT software (relative intensity over 100 units, at the range of 150-600 bp).

AFLP Fragment Recovery and Cloning

AFLP amplified samples were loaded (8 µl) onto a 4.5% polyacrylamide (PAGE) sequencing gel (Sequi-Gen GT, Biorad) and run for 1.5 h at 100W (2090 V - 48 mA), followed by silver staining (Promega, Q4132). AFLP fragments recovered from gels were either reamplified (using 1 µl in PCR reaction with non-labeled primer pair *Eco*-AGT - *Mse*-CAC and *Eco*-AGT - *Mse*-CAA), or cloned into pGEM-T Easy Vector System II and transformed into a Jm109 competent cell (Promega, A1380). Inserts were cut out by *Eco*RI.

ISSR Analysis

For ISSR analysis, the basic protocol of Zietkiewicz *et al.* (1994) was applied using nine primers and combinations: (1) FV808 - $(ag)_8c$; (2) FV810 - $(ga)_8t$; (3) FV811 - $(ga)_8c$; (4) FV819 - $(gt)_8a$; (5) FV820 - $(gt)_8c$; (6) FV821 - $(gt)_8t$; (7) FV835 - $(ag)_8(t/c)c$; (8) FV836 - $(ag)_8(t/c)g$ and (9) FV841 - $(ga)_8(t/c)c$ according to (Cekic *et al.* 2001).

mtDNA

Mitochondrial-DNA (mtDNA) at the 18S-5S rDNA locus was amplified by primer pair: gtg ttg ctg aga cat gcg cc and ata tgg cgc aag acg att cc according to Petit *et al.* (1998).

CAP Analsis of mtDNA

For CAP (*c*leaved *a*mplified *p*olymorphic DNA) analysis, mtDNA fragments amplified at 5S-18S rDNA locus were purified from the EtBr-stained (0.5 ng/µl) agarose (0.8%) gel by a spin column (Sigma G-2166) and digested with seven restriction endonucleases (REs): *Taq*I (T↓CGA), *Bsu*RI (GG↓CC), *Hinf*I (G↓ANTC), *Mbo*I (↓GATC), *Alu*I (AG↓CT) and *Rsa*I (GT↓AC) (2 µl) following the manufacturer's protocol (Fermentas, ER0811). Digested DNA samples (14 µl) were separated on 1.6% agarose gel, and photographed under a transilluminator. *Mbo*I restriction fragments of the medieval sample and current cv '*Omskoje*' (# 13) was purified and sequenced.

Sequencing

Amplified fragments were isolated from the agarose gel with a spin column (Sigma, 56501) and subjected to automated fluorescent DNA sequencing (ABI PRISM 3100 Genetic Analyzer, Applied Biosystems, Hungary). Authenticity of aDNA sequences were proved by BLAST analysis in each case.

Data Analysis

Sequence alignments were analyzed by BioEdit Sequence Alignment Editor (North Carolina State University, USA) (Hall *et al.* 1999), MULTALINE (Combet *et al.* 2000) and CLUSTAL W (Thompson *et al.* 1994) programs. For SSR repeat search FastPCR Program was used (Kalendar *et al.* 2009). BLAST (*B*asic *L*ocal *A*lignment *S*earch *T*ool) analysis was carried out by computer program of NCBI (*N*ational *C*enter for *B*iotechnology *I*nformation, Bethesda, Maryland, USA) (Altschul *et al.* 1997).

CONCLUSION

Seed Samples

Excavation sites from the 4th CENT (Mongolia) and 15th CENT (Hungary) revealed a great number of plant remains, including grains of common millet (*P. miliaceum*) (Figure 1. *a*). The Medieval site (Hungary) revealed about 3 million plant remains of 195 species (Gyulai *et al.* 2006), including 955,497 grains of common millet (Figure 1. b).

Intact grains from both sites after processing were separated from damaged remains under a microscope. The 4th CENT grains showed the highest rate of damage (Figure 1. *a*). After surface sterilization, aseptic seeds were separated and incubated individually (Figure 1. *c*). 16 intact grains from 50 seed remains of 4th CENT millet and 78 intact grains from 150 seed remains of 15th CENT millet were grouped into pooled DNA samples.

Common millet grains of the 15th CENT site appeared to be extremely well preserved due to anaerobic conditions in the slime of a deep well covered by water, apparently used as dust-well in the Middle Ages (Gyulai *et al.* 2006). These seeds looked so intact that we tried to germinate them but with no success (Figure 1. *c*).

Several extinct varieties of crops have been re-introduced into breeding programs by the germination of aged seeds such as a 127-year-old hexaploid Hungarian wheat (Székesfehérvári – Stuhlweissenburger) (*Triticum vulgare* var. *erythrospermum* Körn.) recovered from 1877 in Vienna (Ruckenbauer 1971), a 172-year-old barley (*Hordeum*) and oat (*Avena*) recovered in Nürnberg (Aufhammer and Fischbeck 1964), and the questionable Egyptian wheat *Kamut* (*T. turanicum*) (Quinn 1999) the 10,000-yr-old *Lupinus arcticus* (Porsild *et al.* 1967). None of these programs included DNA analysis.

Despite of our unsuccessful experiments (Figure 1. *c*), aseptic archaeo seeds, free of foreign DNA contaminations, were obtained for further aDNA extraction and molecular analysis.

Quantity and Quality of aDNAs

Agarose gel electrophoresis of the extracted aDNA samples showed different levels of degradation in the 4th and 15th CENT millets compared to current millets, (Figure 1. *d*) probably due to the physical fragmentation and hydrolytic and oxidative damage (Pääbo *et al.* 2004; Schlumbaum *et al.* 2008). The quantity and quality of aDNA extracted from the 4th

CENT and those of 15th CENT millets were significantly less than the twenty DNA samples from current common millets (Figure 2).

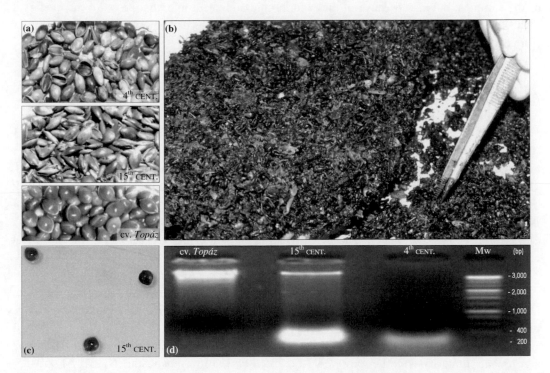

Figure 1. Seed remains of common millet (*P. miliaceum*) recovered from the 4th CENT (Darhan, Mongolia) and 15th CENT (King's Palace of Árpád Dynasty, Budapest, Hungary) compared to current variety cv *Topáz (a)*. Clumps of excavated 15th CENT seeds after wet sieving and floatation *(b)*. Surface sterilized 15th CENT seeds rehydrated on aseptic medium in a germination trial *(c)*. Agarose gel electrophoresis of the extracted aDNAs compared to DNA of current millet cv *Topáz (d)*.

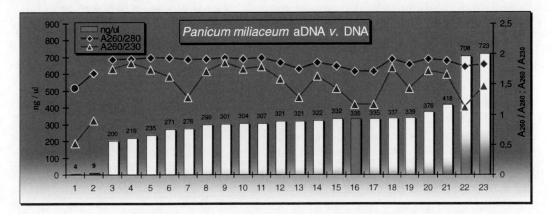

Figure 2. Quantity (0 - 900 ng/μl) and quality (0 – 2.5 ratios of UV Absorbances at 230, 260 and 280 nm) of single seed aDNAs of *(1)* 4th CENT (Darhan, Mongolia) and *(2)* 15th CENT (Budapest, Hungray) millets compared to twenty current common millet (*Panicum miliaceum*) cultivars *(3 - 23)*.

The quality of 15th CENT aDNA showed less degradation with the presence of high molecular weight fragments (Figure 1. *d* and 2); probably not only because of the younger age of the samples, but also because of the cold, humid and anaerobic conditions in the medieval well, similar to aDNA extractions from 400,000 to 10,000-yr-old permafrost plant samples from Siberia (Willerslev *et al.* 2003).

AFLP Loci

Eleven of the 24 selective primer pairs produced significant AFLP patterns in the aDNA samples of millets. The 4th CENT sample revealed two fragments (85 bp and 230 bp) (1.2 %) that had such low intensity (Figure 3a. *A*) that further fragment purification did not succeed. 158 AFLP bands (40.0 %) were amplified in the 15th CENT sample compared to the 264 AFLP bands (100.0 %) detected in the current common millet cv *Topáz*. The average number of AFLP fragments per selective primer pair ranged from 0.2 (4th CENT millet) and 14.4 (15th CENT millet) to 24.0 (*Topáz*). The selective primer pair *Mse*CAA-*Eco*AGT was the most effective combination generating 34 AFLP fragments in the 15th CENT millet (Figure 3a. *B*).

Twenty-one AFLP fragments were recovered from the PAGE gel and eight of them were sequenced after either reamplification or cloning. BLAST analysis revealed that one fragment (*Eco*AGT-*Mse*CAC-272) from the 15th CENT millet showed significant similarity with the universal Ugpe, an ABC-type transporter (permease) gene.

An AFLP fragment (EcoAGT-MseCAA-462) from the current millet cv *Topáz* showed a significant similarity with the gypsy/Ty3-type *Zea mays* retrotransposon (NCBI, AF050455) (Figure 3b) (Gyulai *et al.* 2006).

Microsatellite Loci

SSR alleles at four loci *gln*4, *sh*1, *rps*28 and *rps*15 were amplified in the 15th CENT and current millet cv *Topáz*. No allele was amplified in the 4th CENT sample. After sequencing, the medieval SSR fragments showed identical alignments with current millet sequences (Figure 4).

In theory, microsatellites, as highly species-specific probes, are optimal for aDNA analysis by excluding cross-reactions with contaminating microorganisms, or any ancient organisms and laboratory DNAs (Gugerli *et al.* 2004; Gyulai *et al.* 2006). Unlike wheat, maize, and *Lolium,* there is no database available for common millet. Therefore, we selected maize-specific nuclear SSR primers for DNA-fishing in the aDNA microsatellites. At the *rps*28 locus, only one nucleotide change (A to G) was observed between 15th CENT millet and current millet cv *Topáz* at the 6th position (Figure 4).

As the A to G transition is located in the middle of the SSR sequence, it does not seem to be an artifact caused by *Taq* polymerase which adds an A to the PCR amplified fragments at the 3' end of the sequence (Hofreiter *et al.* 2001).

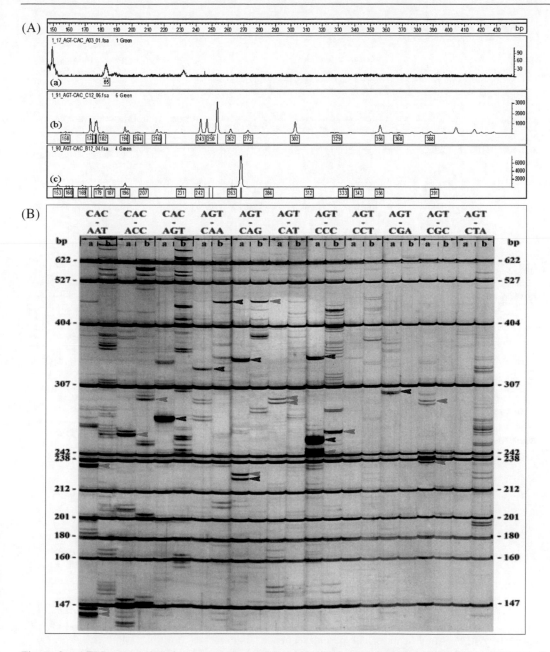

Figure 3a. AFLP. *(A)* AFLP fragments (*Eco*-AGT - *Mse*-CAC) separated by CGE (150 - 430 bp at 0-6000 rel. int.) of the aDNAs of *(a)* 4th CENT and *(b)* 15th CENT millets; and compared to *(c)* a current millet (*Panicum miliaceum*) cv *Topáz*. *(B)* PAGE (4.8 %) separation of eleven AFLP patterns (primer pair sequences are indicated) in the *(a)* 15th CENT and *(b)* current variety cv *Topáz* (DNA Mw markers 622 to 147 bp). Fragments were recovered (green arrowheads) and sequenced (red arrowheads).

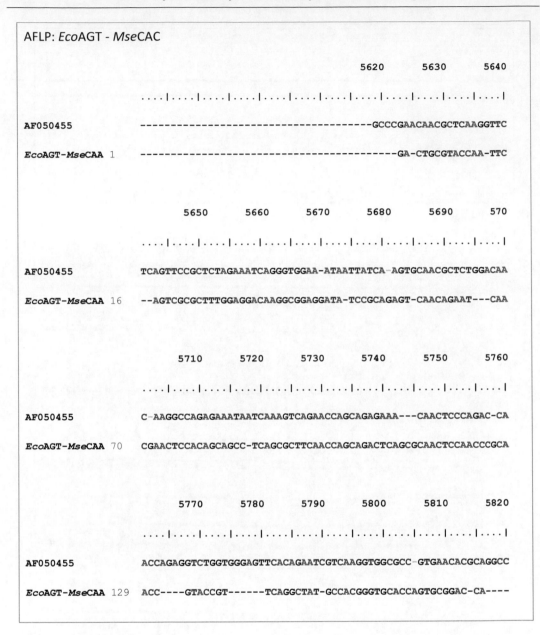

Figure 3b. AFLP. Amplification and partial sequence alignmnets (BioEdit, Hall *et al.* 1999) of a gypsy/Ty3-type retrotransposon from millet DNA (*Panicum miliaceum*) cv *Topáz* amplified by AFLP primer pair *Eco*AGT-*Mse*CAC and compared to sequence of gypsy/Ty3-type *Zea mays* retrotransposon (NCBI, AF050455).

SSR analysis was also applied to herbarium samples of a common reed (*Phargmites australis*) of about 100-years-old to the track plant invasion in North America (Saltonstall 2003). Melon (*Cucumis melo*) specific SSRs were used to identify an *inodorus-* type melon recovered from the 15th CENT (Szabó *et al.* 2005). Allelic diversity of microsatellites were also reliably detected in aDNAs of 4000-yr-old seagrass (*Posidonia oceanica*) (Raniello and Procaccini 2002).

SSRs

(1) *rps*28

```
                    280        290        300        310        320        330        340        349
                     |          |          |          |          |          |          |          |
▶ 15th cent. millet  TCACG GCTTGGAATGAAGGGAG CTCCTCTCT----CTCTCTCTCT CATTTCCATAATTTATATATTGTGTGTAGATAG
P.miliaceum cv Topáz ..... ................ CTCCTCTCT----CTCTCTCTCT ................................
Zea mays, EU9687241  C.T.. .......T........ CTCCTCTCTGTCTCTCTCTCTCT ..........A.........T.....
```

(2) *rps*15

```
                      2090       2100       2110       2120       2130       2140       2150       2160       2170       2180
▶ 15th cent. millet  ----GGGAGGGCCATGACCATGGATGATGGTGACGGCCACAGCAGCAGCAGCAGCGACAGCGACTGATCTCGCCTCGCCGAGCGCCGTACGCACAAATGC
P.m.cv. Topáz        ----GGGAGGGCCATGACCATGGATGATGGTGACGGCCACAGCAGCAGCAGCAGCGACAGCGACTGATCTCGCCTCGCCGAGCGCCGTACGCACAAATGC
EU9884481.Z.mays     CAACGGCAAGACCTTCAACCAGGTTGAGATCAAGCCTGAGATGATCGGCCACTACCTTGCCGAGTTCTCCATCTCCTATAAGCCGGTCAAGCATGGCAGG
Z231621. A.th.       CAACGGAAAGACCTTCAACCAGGTTGAGATCAAGCCTGAGATGATTGGTCACTACCTGGCTGAGTTCTCTATCTCATACAAGCCGGTCAAGCACGGTAGG
```

(3) *sh*1

```
                        7690       7700       7710       7720       7730       7740       7750       7760
                         |          |          |          |          |          |          |          |
▶ 15th cent. millet     GCCGTTTGATGCCTGCCTCTGCCTCTGCCTGCCTGCCAGAGAGGATCACGTGTTC-----GTTCCCATTCCAGCAGTC
P. miliaceum cv. Topáz  GCCGTTTGATGCCTGCCTCTGCCTCTGCCTGCCTGCCAGAGAGGATCACGTGTTC-----GTTCCCATTCCAGCAGTC
AF5441151.Zea mays      GCCGTTTGATGCCTGCCTCTGCCTCTGCCTGCCTGCCAGAGAGGATCACGTGTTC-----GTTCCCATTCCAGCAGTC
AY1182661.Saccharum o.  GTTTTTTGATGTCT-------------------TGCTAGAGAGAAATATGTGTAGCATCAAGTGTTCAGTTCAGCAGTC
FJ5133251.Sorghum bicolor GTTATGATGT----------------------CTGCTAGAGAGAACGTGTAGCATGCAGTGTTCATTTTTCAGCAGTG

                        7770       7780       7790       7800       7810       7820       7830       7840
                         |          |          |          |          |          |          |          |
                        TTAAACGAGTGAATGAACTACTGACGCATCTCTTTCTGGAATCCAGGCAAGCCAGGTTCCGCTGTCCTTCGATTAGTACG
                        TTAAACGAGTGAATGAACTACTGACGCATCTCTTTCTGGAATCCAGGCAAGCCAGGTTCCGCTGTCCTTCGATTAGTACG
                        TTAAACGAGTGAATGAACTACTGACGCATCTCTTTCTGGAATCCAGGCAAGCCAGGTTCCGCTGTCCTTCGATTAGTACG
                        TTAAACAAGTGGATTAACTAGTA---CATCTCTTTCTTGAATCCAGGCAAGTGCGGTTCCATTGTGCCTCGATTAGTGTG
                        TTAAACAAGCGGATTTACTAATG---CATCTCTTTCTTGAATCCAGGCAAGCGCGGTTCCCTTGTCCTACGACTAGTGTG
```

(4) *gln*4

```
                          1050       1060       1070       1080       1090       1100       1110       1120
                           |          |          |          |          |          |          |          |
▶ 15th cent. millet                                   TCGAGGACCGCCG-CCGGCGTCCA-CATGGACCCCTACGTGGTCACCTCC
P. miliaceum cv. Topáz                                TCGAGGACCGCCG-CCGGCGTCCA-CATGGACCCCTACGTGGTCACCTCC
NM_0011118261.Zea mays       GGGAGACGGAGCAGAACGGCAAGGGCTACTTCGAGGACCGCCGCCGGCGTCCAACATGGACCCCTACGTGGTCACCTCC
DQ1242111.Triticum aestivum  GGGAGACGGAGCAGAACGGCAAGGGCTACTTCGAGGACCGCCGGCCGGCGTCCAACATGGACCCCTACGTGGTCACCTCC
X690871.Horedum vulgare      GGGAGACGGAGCAGAACGGCAAGGGCTACTTCGAGGACCGCCGGCCGGCGTCCAACATGGACCCCTACGTGGTCACCTCC
AB0375951.Oryza sativa       GGGAGACGGAGCAGAACGGCAAGGGCTACTTCGAGGATCGCCGGCCGGCGTCCAACATGGACCCTTACATCGTCACCTCC
AY8354531.Saccharum o.       GGGAGACAGAGCAGAACGGCAAGGGCTACTTCGAGGACCGCCGGCCGGCGTCCAACATGGACCCCTACGTGGTCACCTCC

                          1130       1140       1150       1160       1170       1180       1190       1200
                           |          |          |          |          |          |          |          |
                          ATGATCGCCGAGTACACCATCATCTGGA-GCCCTGAGGGCTA----------------AGGCGGCCGTTCCGTCGCGT
                          ATGATCGCCGAGTACACCATCATCTGGA-GCCCTGAGGGCTA----------------AGGCGGCCGTTCCGTCGCGT
                          ATGATCGCCGAGACCACCATCATCTGGAAGCCCTGAGGGCTA----------------AGGCGGCCGTTCCGTCGCGT
                          ATGATCGCCGAGACCACCATCCTGTGGAAGCCCTGAGGCAGCGC---------------------CGACCGCCG
                          ATGATCGCCCAGACCACCATCCTGTGGAAGCCCTGAAGCT--------C-----------------CGATCGCCG
                          ATGATCGCCGAGACCACCATCATCTGGAAGCCCTGAAGCGGCT---------------------------
                          ATGATCGCCGACACCACCATCCTGTGGAAGCCCTGAGGCGTCCAAGCTTTTCCTCCATCCAGCTCGCCGTGTCGTGTCGG
```

Figure 4. SSR. Consensus sequence alignments (MULTALINE, Combet *et al.* 2000) of partial SSR loci of *(1) rps*28, *(2) rps*15, *(3) sh*1, and *(4) gln*4 amplified from the aDNA of 15th CENT millet (▶) and compared to a current millet (*Panicum miliaceum*) cv *Topáz* with a single SNP (G to A at 6th nt) at the *(1) rps*28 site. (CT repeats in frames with NCBI accessions of monocots).

ISSR Loci

Seven of the nine ISSR primers amplified fragments in the medieval and current millets and only two FV819 - (gt)$_8$a and FV820 - (gt)$_8$c failed amplification (Lágler *et al.* 2005, 2006). In total, 15 ISSR alleles were detected in the 15th CENT millet. A molecular dendrogram based on the presence versus absence of ISSR fragments (Figure 5a) indicated that medieval common millets showed the closest genetic similarity to a current variety

'Omskoje' (# 13) (Lágler *et al.* 2005, 2007) (Figure 5c and 7). Unexpectedly, no SNPs were detected in the sequences of ISSR fragments (Figure 5b), which might indicate the applicability of ISSR for the detection of conserved DNA regions (Zietkiewicz *et al.* 1994). Dominant markers, such as ISSR, are generally less suitable for genotyping aDNA because the absence of a fragment could be either due to the loss of the degraded DNA sequence or the variation of that locus in the genome (Biss *et al.* 2003). However, ISSR was very useful for sequence amplification from aDNA samples.

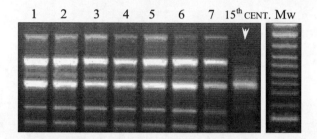

Figure 5a. ISSR. Loss of ISSR loci in the 15th CENT millet (arrow) compared to seven current millet cultivars (*Panicum miliaceum*) on agarose gel (0.8 %) amplified by primer $(GA)_8C$ (Cekic *et al.* 2001) (Mw 100 bp DNA ladder).

```
ISSR: (GA)₈T - (GA)₈(T/C)C

                         10        20        30        40        50        60        70        80
                          |         |         |         |         |         |         |         |
► 15th cent.millet    TAT●TCTCTTTACCGCTACCTACCGCTCCTACTATTTCGTTGTCAGTCCCTGCCCCTGTCCTTCAGAGAGGAAAAGCGAC
P.m.cv.Omskoje        TATCTCTCTTTACCGCTACCTACCGCTCCTACTATTTCGTTGTCAGTCCCTGCCCCTGTCCTTCAGAGAGGAAAAGCGAC
P.m.cv.Csaszarreti    TATCTCTCTTTACCGCTACCTACCGCTCCTACTATTTCGTTGTCAGTCCCTGCCCCTGTCCTTCAGAGAGGAAAAGCGAC
P.m.cv.Jaszberenyi    TATCTCTCTTTACCGCTACCTACCGCTCCTACTATTTCGTTGTCAGTCCCTGCCCCTGTCCTTCAGAGAGGAAAAGCGAC
P.m.cv.Kecskemeti     TATCTCTCTTTACCGCTACCTACCGCTCCTACTATTTCGTTGTCAGTCCCTGCCCCTGTCCTTCAGAGAGGAAAAGCGAC
P.m.cv.Fertodip.      TATCTCTCTTTACCGCTACCTACCGCTCCTACTATTTCGTTGTCAGTCCCTGCCCCTGTCCTTCAGAGAGGAAAAGCGAC

                         90       100       110       120       130       140       150       160
                          |         |         |         |         |         |         |         |
                      GATAAACAACCAATGTCGGAATGTCCAGAACCTCTGAACAAAAAATTAATTGCCCAAGCTCGCTCTAGACCACTTTAGTT
                      GATAAACAACCAATGTCGGAATGTCCAGAACCTCTGAACAAAAAATTAATTGCCCAAGCTCGCTCTAGACCACTTTAGTT
                      GATAAACAACCAATGTCGGAATGTCCAGAACCTCTGAACAAAAAATTAATTGCCCAAGCTCGCTCTAGACCACTTTAGTT
                      GATAAACAACCAATGTCGGAATGTCCAGAACCTCTGAACAAAAAATTAATTGCCCAAGCTCGCTCTAGACCACTTTAGTT
                      GATAAACAACCAATGTCGGAATGTCCAGAACCTCTGAACAAAAAATTAATTGCCCAAGCTCGCTCTAGACCACTTTAGTT
                      GATAAACAACCAATGTCGGAATGTCCAGAACCTCTGAACAAAAAATTAATTGCCCAAGCTCGCTCTAGACCACTTTAGTT

                        170       180       190       200       210       220       230       240
                          |         |         |         |         |         |         |         |
                      TTACATCCGGGGTGCCACTCAAAACCCTTCATTAATTCTACCTGCCGACTCTAACCAATGTCGGAACGTCCAAAGACTCT
                      TTACATCCGGGGTGCCACTCAAAACCCTTCATTAATTCTACCTGCCGACTCTAACCAATGTCGGAACGTCCAAAGACTCT
                      TTACATCCGGGGTGCCACTCAAAACCCTTCATTAATTCTACCTGCCGACTCTAACCAATGTCGGAACGTCCAAAGACTCT
                      TTACATCCGGGGTGCCACTCAAAACCCTTCATTAATTCTACCTGCCGACTCTAACCAATGTCGGAACGTCCAAAGACTCT
                      TTACATCCGGGGTGCCACTCAAAACCCTTCATTAATTCTACCTGCCGACTCTAACCAATGTCGGAACGTCCAAAGACTCT
                      TTACATCCGGGGTGCCACTCAAAACCCTTCATTAATTCTACCTGCCGACTCTAACCAATGTCGGAACGTCCAAAGACTCT
```

Figure 5b. ISSR. Consensus sequence alignments (MULTALINE, Combet *et al.* 2000) of ISSR fragments (1 to 240 bp) of the 15th CENT millet (►) and compared to five current (*Panicum miliaceum*) cultivars amplified by primer pair $(GA)_8T$ and $(GA)_8(T/C)C$ (Cekic *et al.* 2001) with complete sequence homology of five to six nucleotide motives.

RAPD polymorphism clustered common millet accessions according to geographical regions of origin, which indicated that the genome of common millet was fairly plastic with the potential for a relatively rapid adaptation period during microevolution (M'Ribu and Hilu 1994).

Isozyme (Warwick 1987) and RFLP (Parani *et al.* 2001) and analyses also revealed significant molecular differences among millet accessions. An olive-black seeded biotype called 'wild proso millet', which appeared in North America 1970, after about 400 years of millet cultivation (as millet was introduced to North America in the 17th century) also indicated a plastic common millet genome (Strand and Behrens 1981). New, seedy biotypes of common millet were also reported from Europe (Scholz and Mikolas 1991).

CAP-mtDNA at 5S-18S rDNA locus

The mitochondrial specific primer pair amplified an 1117 bp fragment at the 5S-18S rDNA locus of mtDNA in 15th CENT millet and current cv '*Omskoje*' (# 13). DNA fragments were digested at six restriction endonucleases (*Taq*I, *Bsu*RI, *Hinf*I, *Mbo*I, *Alu*I and *Rsa*I) for CAP analysis.

The pattern of digested fragments of *Mbo*I with four digested (44-, 91-, 193-, 92 and 572 bp) fragments showed the same sizes and numbers, which indicated that no nucleotide substitutions occurred in the restriction sites (Figure 6).

mtDNA: *Mbo*I

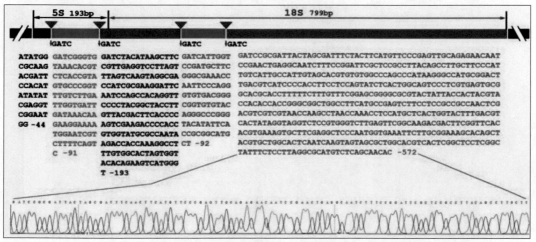

Figure 6. mtDNA. Sequence analysis of *Mbo*I digested (↓GATC) aDNA fragment of 15th CENT millet at the 5S (193 bp) - 18S (799 bp) rDNA locus of mtDNA with partial fluorogram of the 572 bp fragment.

Sequence analysis of the *Mbo*I restricted fragments (1117 bp) revealed consensus sequence alignments (Figure 6) without SNPs due to the highly conserved nucleotide sequence of mitochondrial DNA (Petit *et al.* 1998). Multiple copy DNAs, as the highly conserved mtDNA, cpDNA and ribosomal rDNA, were also reported to yield reliable amplification compared to double-copy (2n) nuclear aDNA sequences (Deguilloux *et al.* 2002; Gugerli *et al.* 2005). No organelle aDNA was amplified in the 4th CENT millet. All of this data indicates that due to the highly conserved sequences detected at mtDNA and ISSR loci and the relatively short time of 600 years for microevolution no mutations of SNPs were detected.

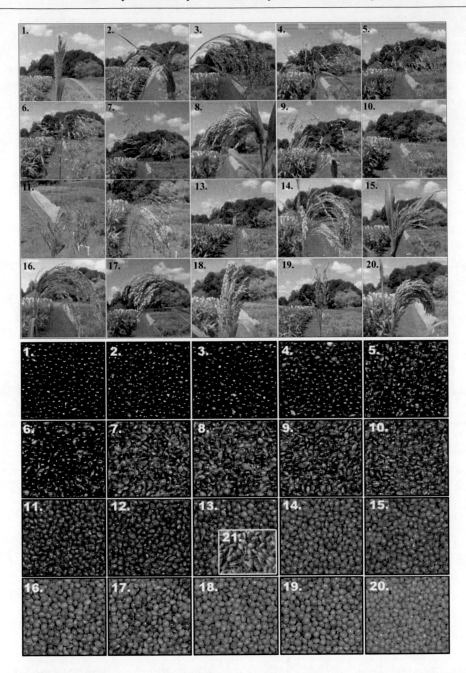

Figure 7. Types of panicles *(up)* and grains *(down)* of current millet cvs. (*Panicum miliaceum*) (# 1 - 20) used for phenotype reconstruction of 15th CENT millet (# 21).

To conclude, the incubation of archaeobotanical seed samples in aseptic condition provides contamination free aDNA samples. aDNAs can be analyzed successfully using different techniques of AFLP, SSR, ISSR, CAP for nuclear DNA and for organelle mtDNA. The two AFLP fragments of the 4th CENT indicated the time limitations of archaeogentics (Willerslev *et al.* 2003). Sequence analysis of the aDNA fragments of 15th CENT millet showed identical common millet origin with only one SNP at the 6th position (A to G) at the *rps*28 locus. All sequencing data of SSRs, mtDNA loci and ISSR fragments pattern indicated a conserved *Panicum* genome. Comparative analysis, based on ISSR fragment diversity revealed that 15th CENT medieval millet showed the closest genetic similarity to a registered current cultivar #13 '*Omskoje*' of ancient characters of 'light' panicle type with small grains (Accession # 02546/00, ABI Tápiószele, Hungary, H-2766) (Figure 7), which might provide an insight to reconstruct the medieval common millet grown in the 15th century Hungary.

REFERENCES

Altschul SF, TL Madden, AA Schaffer, JH Zhang, Z Zhang, W Miller and DJ Lipmand (1997) Gapped BLAST and PSI-BLAST: a new generation of protein database search programs. *Nucleic Acids Research* 25: 3389-3402.

Aufhammer G, G Fischbeck (1964) Ergebnisse von Gefass- und Feldversuchen mit dem Nachbau keimfahiger Gersten- und Haferkorner aus dem Grundstein des 1832 errichteten. Nürnberger Stadttheates. *Z Pflanzenzuchtung* 51: 345-378.

Baltensperger DD, LA Nelson, GE Frickel, RL Anderson (1997) Registration of 'Sunrise' proso millet. *Crop Sci* 37: 1380.

Biss P, J Freeland, J Silvertown, K McConway, P Lutman (2003) Successful amplification of rice chloroplast microsatellites from century-old grass samples from the park grass experiment. *Plant Mol Biol Rep* 21: 249-257.

Bisztray GyD, R Bacsó, P Bodor, G Facsar, F Gyulai, I Velich (2004) Archaeobotanical and genetical methods to analyse 600-years-old seeds of horticultural plants. Book of Abstracts 5th IVCHB Symposium, In Vitro Culture and HorticulturaBreeding,12-17. September 2004, Debrecen, Hungary, p. 212.

Cekic C, NH Battey, MJ Wilkinson (2001) The potential of ISSR-PCR primer-pair combinations for genetic linkage analysis using the seasonal flowering locus in *Fragaria* as a model. *Theor Appl Genet* 103: 540-546.

Chin E, L Senior, H Shu, JSC Smith (1996) Maize simple repetitive DNA sequences: Abundance and allele variation. *Genome* 39: 866-873.

Colosi JC, BA Schaal (1997) Wild proso millet (*Panicum miliaceum*) is genetically variable and distinct from crop varieties of proso millet. *Weed Science* 45: 509-518.

Combet C, C Blanchet, C Geourjon, G Deléage (2000) NPS@: Network Protein Sequence Analysis. *Trends in Biochemical Sciences* 25(3): 147-150.

Cresswell A, NR Sackville-Hamilton, AK Roy, BMF Viegas (2001) Use of AFLP markers to assess genetic diversity of *Lolium* species from Portugal. *Molecular Ecology* 10: 229-241.

Deguilloux M-F, M-H Pemonge, RJ Petit (2002) Novel perspectives in wood certification and forensics: dry wood as a source of DNA. *Proceedings of the Royal Society of London, Series B* 269: 1039-1046.

Doyle JJ, JL Doyle (1990) Isolation of plant DNA from fresh tissue. *Focus* 12: 13-15.

Gorman CF (1969) Hoabinhian: A pebble-tool complex with early plant associations in Southeast Asia. *Science* 163: 671-673.

Gugerli F, L Parducci, RJ Petit (2005) Ancient plant DNA: review and Prospects. *New Phytologist* 166: 409-418.

Gyulai G, J Janovszky, E Kiss, L Lelik, A Csillag, LE Heszky (1992) Callus initiation and plant regeneration from inflorescence primordia of the intergeneric hybrid *Agropyron repens* (L.) Beauv. x *Bromus inermis* Leyss. cv. *nanus* on a modified nutritive medium. *Plant Cell Rep* 11: 266-269.

Gyulai G, A Magda, J Kiss, F Gyulai, L Holly, L Heszky (2001) DNS izolálás és PCR-amplifikáció 700 éves növény magvakból. VII. Növénynemesítési Tudományos Napok, Budapest, p.89.

Gyulai G, Z Mester, J Kiss, L Szemán, L Heszky, A Idnurm (2003) Somaclone breeding of reed canarygrass (*Phalaris arundinacea* L). *Grass and Forage Science* 58: 210-215.

Gyulai G, M Humphreys, A Bittsánszky, K Skøt, J Kiss, L Skøt, G Gullner, S Heywood, Z Szabó, A Lovatt, L Radimszky, H Roderick, M Abberton, H Rennenberg, T Kőmíves, L Heszky (2005) AFLP analysis and improved phytoextraction capacity of transgenic *gsh*I-poplar clones (*Populus canescens* L.) in vitro. *Zeitschrift für Naturfroschung* 60c: 523-537.

Gyulai G, M Humphreys, R Lágler, Z Szabó, Z Tóth, A Bittsánszky, F Gyulai, L Heszky (2006) Seed remains of common millet from the 4th (Mongolia) and 15th (Hungary) centuries: AFLP, SSR and mtDNA sequence recoveries. *Seed Science Research* 16: 179-191.

Hall TA (1999) BioEdit: a user-friendly biological sequence alignment editor and analysis program for Windows 95/98/NT. *Nucleuc Acids Symposium Ser.*: 41: 95-98.

Harlan JR (1971) Agricultural origins: centers and noncenters. *Science* 174: 468-473.

Ho P-t (1977) The indigenous origins of Chinese agriculture. In: *Origins of Agriculture* (Ed) Reed CA, pp. 413-418. Mouton, Publ., Paris.

Hodkinson TR, SA Renvoize, GN Chonghaile, C Stapleton, MW Chase (2000) A comparison of ITS nuclear rDNA sequence data and AFLP markers for phylogenetic studies in Phyllostachys (Bambusoideae, Poaceae). *J Plant Res* 113: 259-269.

Hofreiter M, V Jaenicke, D Serre, A von Haeseler, S Pääbo (2001) DNA sequences from multiple amplifications reveal artifacts induced by cytosine deamination in ancient DNA. *Nucleic Acids Research* 29: 4793-4799.

Kalendar R, D Lee, AH Schulman (2009) FastPCR Software for PCR Primer and Probe Design and Repeat Search. *Genes, Genomes and Genomics*, 3(1): 1-14. [http://primerdigital.com/fastpcr.html].

Keng H (1974) Economic plants of ancient north China as mentioned in *Shih Ching* (Book of Poetry). *Economic Botany* 28: 391-410.

Lágler R, G Gyulai, M Humphreys, Z Szabó, L Horváth, A Bittsánszky, J Kiss, L Holly, L Heszky (2005) Morphological and molecular analysis of common millet (*P. miliaceum*) cultivars compared to an aDNA sample from the 15th century (Hungary). *Euphytica* 146: 77-85.

Lágler R, G Gyulai, Z Szabó, Z Tóth, A Bittsánszky, L Horváth, J Kiss, F Gyulai, L Heszky (2006) Molecular diversity of common millet (*P. miliaceum*) compared to archaeological samples excavated from the 4th and 15th centuries. *Hung Agric Res* 2006/1:14-19.

Lágler R, Gyulai G, Szabó Z, Tóth Z, Heszky L (2007) A köles (*Panicum miliaceum*) SSR-és ISSR szekvencia-stabilitása a 4. és 15. századi régészeti leletektől a mai fajtákig. *Agrártud Közl* 27: 10-19.

M'Ribu HK, KW Hilu (1994) Detection of interspecific and intraspecific variation in *Panicum* millets through random amplified polymorphic DNA. *Theor Appl Genet* 88: 412-416.

Murray MG, WF Thompson (1980) Rapid isolation of high molecular weight plant DNA. *Nucleic Acids Res* 8: 4321-4325.

Nadel D, U Grinberg, E Boaretto, E Werker (2006) Wooden objects from Ohalo II (23,000 cal BP), Jordan Valley, Israel. *J Human Evol* 50: 644-662.

Pääbo S, H Poinar, D Serre, V Jaenicke-Despres, J Hebler, N Rohland, M Kuch, J Krause, L Vigilant, M Hofreiter (2004) Genetic analyses from ancient DNA. *Annual Review of Genetics* 38: 645-679.

Parani M, K Rajesh, M Lakshmi, L Parducci, AE Szmidt, A Parida (2001) Specific identification in seven small millet species using polymerase chain reaction – restriction fragment length polymorphism of trnS-pbsC gene region. *Genome* 44: 495-499.

Petit RJ, B Demesure, S Dumolin (1998) cpDNA and mtDNA primers in plants. In: A Karp, P.G. Isaac, D.S. Ingram (Ed) Molecular tools for screening biodiversity. pp. 256-261. Chapman and Hall, London, ISBN 0 412 63830 4.

Porsild AE, CR Pharington, GA Mulligan (1967) '*Lupinus arcticus*'. Whats grown from seeds of Pleistocene age. *Science* 158: 113-114.

Quinn RM (1999) Kamut: Ancient grain, new cereal. In: J. Janick (Ed.) *Perspectives on new crops and new uses*. pp. 182 -183, ASHS Press, Alexandria, VA.

Raniello R, G Procaccini (2002) Ancient DNA in the seagrass *Posidonia oceanica*. *Marine Ecology – Progress Series* 227: 269-273.

Ruckenbauer P von (1971) Keimfähiger Winterweizen aus dem Jahre 1877. Beobachtungen und Versuche. pp. 372-386. Inst. f. Pflanzenbau und Pflanzenzuchtung d. *Hochschule f. Bodenkultur in Wien*.

Saltonstall K (2003) Microsatellite variation within and among North American lineages of *Phragmites australis*. *Molecular Ecology* 12: 1689-1702.

Schlumbaum A, M Tensen, V Jaenicke-Déprés (2008) Ancient plant DNA in archaeobotany. *Veget Hist Archaeobot* 17: 233-244.

Scholz H, V Mikolas (1991) The weedy representatives of proso millet (*Panicum miliaceum, Poaceae*) in Central Europe. *Thaiszia* 1: 31-41.

Skøt L, NRS Hamilton, S Mizen, KH Chorlton, ID Thomas (2002) Molecular genecology of temperature response in *Lolium perenne*: 2. association of AFLP markers with ecogeography. *Molecular Ecology* 11: 1865-1876.

Smith PM (1977) Minor crops. In: Evolution of crop plants. (Ed) Simmonds NW, pp. 301-324. Longman, London and New York.

Strand OE, R Behrens (1981) Identification of wild proso millet. Minneapolis, NM: Agronomy Fact Sheet No.35, University of Minnesota. 2 p.

Suh HS, JH Cho, YJ Lee, MH Heu (2000) RAPD variation of 13,100 and 17,310 year-old carbonized rice. In: B. Gordon (Ed) 4th International Rice Genetics Symposium, Manilla, Philipines, Oct. 22-27, Manila, Philipines. (without page numbers), (http://www.carleton.ca).

Szabó Z, G Gyulai, M Humphreys, L Horváth, A Bittsánszky, R Lágler, L Heszky (2005) Genetic variation of melon (*C. melo*) compared to an extinct landrace from the Middle Ages (Hungary) I. rDNA, SSR and SNP analysis of 47 cultivars. *Euphytica* 146: 87-94.

Thompson JD, DG Higgins, TJ Gibson (1994) CLUSTAL W: improving the sensitivity of progressive multiple sequence alignment through sequence weighting, positions-specific gap penalties and weight matrix choice. *Nucleic Acids Research* 22: 4673-4680.

Tseveendorj D, L Sugar (1994) Hanguk Minjokhak Yön'gu. In.: *The review of Corean Anthropology Institute* 2, pp. 91-110, Dankook University, S Korea.

Vaughan DA, E Balázs, JS Heslop-Harrison (2007) From Crop Domestication to Super-domestication. *Annals of Botany* 100: 893-901.

Vos P, R Hogers, M Bleeker, M Reijans, T Lee, M Hornes, A Friters, J Pot, J Paleman, Mn Kuiper, M Zabeau (1995) AFLP: a new technique for DNA fingerprinting. *Nucleic Acids Research* 23: 4407-4414.

Walters TW (1989) Historical overview on domesticated plants in China with special emphasis on the *Cucurbitaceae*. *Economic Botany* 43: 297-313.

Warwick SI (1987) Isozyme variation in proso millet. *J Heredity* 78: 210-212.

Willerslev E, AJ Hansen, J Binladen, TB Brand, M Thomas, P Gilbert, B Shapiro, M Bunce, C Wiuf, DA Gilichinsky, A Cooper (2003) Diverse plant and animal genetic records from Holocene and Pleistocene sediments. *Science* 300: 791-795.

Zietkiewicz E, A Rafalski, A Labuda (1994) Genome fingerprinting by simple sequence repeat (SSR)-anchored polymerase chain reaction amplification. *Genomics* 20: 176-183.

In: Plant Archaeogenetics
Editor: Gábor Gyulai, pp. 69-87

ISBN: 978-1-61122-644-7
© 2011 Nova Science Publishers, Inc.

Chapter 7

FLESH COLOR RECONSTRUCTION FROM aDNAs OF *Citrullus* SEEDS FROM THE 13^(TH), 15^(TH), AND 19^(TH) CENTs (Hungary)

Gábor Gyulai[1,], Zoltán Tóth[1] and András Bittsánszky[1,2]*
[1]St. Stephanus University, Institute of Genetics and Biotechnology,
Gödöllő 2103 Hungary
[2]Plant Protection Institute, Hungarian Academy of Sciences,
Budapest 1525 Hungary

ABSTRACT

Seed remains of watermelon (*Citrullus lanatus lanatus*) were excavated from two sites from the Middle Ages 13^(th) (Debrecen) and 15^(th) centuries (Budapest), Hungary, processed by floatation followed by seed sorting and identification. After seed morphological analysis aDNAs were extracted and analyzed at loci of twelve microsatellites (SSRs), two cpDNA, RAPD, ITS, and *lcyb* gene with a final aim of sequence recovery and phenotype reconstruction. For comparative analysis, an herbarium sample from the 19^(th) CENT (Pannonhalma, Hungary) and forty-four current *Citrullus* cultivars were used. Molecular dendrogram based on microsatellite analysis revealed the closest similarity to current watermelon cultivars with red flesh (13^(th) CENT) and yellow flesh (15^(th) CENT) colors. The 170-yr-old herbarium sample showed close molecular similarity to citron melons (*Citrullus lanatus citroides*) which also reflects the importance of citron melon as fodder in Hungary.

* E-mail: gyulai.gabor@mkk.szie.hu

INTRODUCTION

The aDNA (*a*ncient DNA) recovered from excavated remains of plants and animals supply unique materials not only for the analysis of post-mortem DNA degradation (Brown 1999; Threadgold and Brown 2003), but also for tracing vegetation history and microevolution (Gugerli *et al*. 2005; Schlumbaum *et al*. 2007). Intact aDNA sequences (Szabó *et al*. 2005; Lágler *et al*. 2005; Gyulai *et al*. 2008) and complete genome (Cooper *et al*. 2001) of the extinct organisms can be reconstructed in the case of optimal preservation conditions. In this chapter we present the analyses of seed morphology and aDNA study of 700-, 600- and 170-year-old watermelons together with a comparison to current cultivars.

MATERIALS AND METHODS

Seed Samples

Seed remains of watermelon (*Citrullus l. lanatus*) from the 13th CENT were excavated in Debrecen, Hungary. In total, 95,133 seeds of 206 plant species were identified. Of them, 251 watermelon seeds were determined to have the same morphological characters. At the 15th CENT site (8th well, Mansion Teleki, King's Palace of Árpád Dynasty, Buda Hill, Budapest; Hungary) 54,415 watermelon seeds were excavated (Gyulai *et al*. 2008; Tóth *et al*. 2008a,b). Wet-sieved sediment samples were processed by floatation followed by seed sorting and identification in the laboratory (Schermann 1966; Hartyányi and Nováki 1975). The 19th CENT (from ca 1836) seeds were collected from the oldest botanical seed collection of Hungary (Pannonhalma, Hungary) (Vörös 1971), recently exhibited at the Hungarian Agricultural Museum, Budapest (Hungary) (Figure 1). For comparative analysis, forty-four current *Citrullus* species and varieties were used including colocynth (*C. colocynthis*, #1-3), citron melon (*C. l. citroides*, # 4-6) and watermelons (*C. l. lanatus*, # 7-44) (Horváth *et al*. 2007, Meszter 2006).

Elimination of Contaminations

Seeds were incubated for seven days in an aseptic tissue culture medium to eliminate contamination before DNA extraction prior to washing with a detergent (3 min), rinsing three times with distilled water (3 min), surface sterilization with ethanol (70% v/v) using a bleaching agent (8% $Ca(OCl)_2$ w/v) for 1 min, followed by three rinses with sterile distilled water according to general aseptic culture techniques (Gyulai *et al*. 2001, 2008). Seeds of the current varieties were also surface sterilized. Exogenously and endogenously contaminated seeds infected by fungi and bacteria were eliminated from further analyses.

aDNA Extraction

Individual seeds were ground in an aseptic mortar with liquid nitrogen in a laminar air flow cabinet in the archaeobotanical laboratory of the St Stephanus University, Gödöllő. aDNA was extracted by the CTAB (cethyltrimethylammonium bromide) method according to Biss *et al.* (2003), Yang (1997), Cooper and Poinar (2000) and Bisztray *et al.* (2004).

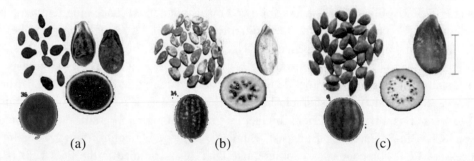

(a)   (b)   (c)

Figure 1. Ancient watermelon (*Citrullus l. lanatus*) seeds (groups and single, bar 1cm) excavated from *(a)* 13th CENT (Debrecen, Hungary) with kernel remains and *(b)* 15th CENT (Budapest, Hungary), and compared to *(c)* a herbarium sample of citron melon (*Citrullus l. citroides*) from ca 1830 (Pannonhalma, Hungary). Reconstructed flesh types based on SSRs (*Figure 5*), and *lcyb* gene sequences (*Figure 8*) are also indicated by the genetically closest current varieties *(a)* #36 *Kecskeméti vöröshéjú, (b)* #14 *Csárdaszállás,* and *(c)* #6 *Újszilvás.*

Seed DNA of current cultivars (0.1 g) was also extracted in CTAB buffer, followed by an RNase-A treatment (Sigma, R-4875) for 30 min at 37 °C in each case. The quality and quantity of extracted DNA was measured (2 µl) by a NanoDrop ND-1000 UV-Vis spectrophotometer (NanoDrop Technologies, Delaware, USA – BioScience, Budapest, Hungary) (Figure 2). DNA samples were adjusted to a concentration of 30 ng/µl with ddH$_2$O and subjected to PCR amplification.

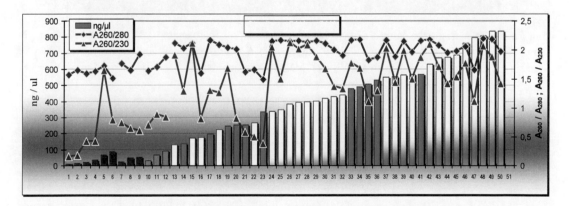

Figure 2. Quantity (0 – 900 ng/µl) and quality (0 – 2.5 ratios of UV Absorbances at 230, 260 and 280 nm) of single seed aDNAs *(1-6)* from 13th CENT *Citrullus l. lanatus*, Debrecen, Hungary, *(7-9)* 15th CENT *Citrullus l. lanatus*, Budapest, Hungary, *(10-12)* 19th CENT *Citrullus l. citroides*, Pannonhalma (Hungary), and compared to *(13-51)* current *Citrullus* cultivars.

aDNA Amplification by PCR

Hot Start PCR (Erlich *et al.* 1991) was combined with Touchdown PCR (Don *et al.* 1991) using AmpliTaq Gold™ Polymerase. Reactions were carried out in a total volume of 25 µl (containing genomic DNA of 30-50 ng, 1 x PCR buffer (2.5 mM $MgCl_2$), dNTPs (200 µM each), 20 pmol of each primer and 1.0 U of *Taq* polymerase. Touchdown PCR was performed by decreasing the annealing temperature by 1.0 °C / per cycle with each of the initial 12 cycles (PE 9700, Applied Biosystems), followed by a 'touchdown' annealing temperature for the remaining 25 cycles at 56 °C for 30 s with a final cycle of 72 °C for 10 min and held at 4 °C. A minimum of three independent DNA preparations of each sample was used. Amplifications were assayed by agarose (0.8 %, SeaKem LE, FMC) gel electrophoresis (Owl system), stained with ethidium bromide (0.5 ng/µl) after running at 80 V in 1 X TBE buffer. Each successful reaction with scorable bands was repeated at least twice. Transilluminated gels were analyzed by the ChemiImager v 5.5 computer program (Alpha Innotech Corporation - Bio-Science Kft, Budapest, Hungary). A negative control which contained all the necessary PCR components except template DNA was included in the PCR runs (Figure 3).

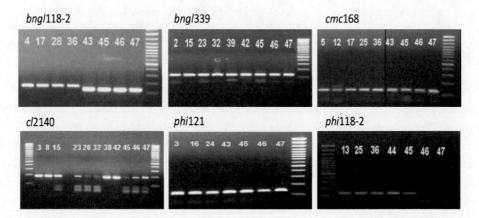

Figure 3. aDNA fragments amplified by PCR and separated on agarose gel (0.8 %) at six aDNA loci in the 13th CENT *(45)*, 15th CENT *(46)* and 19th CENT *(47)* *Citrullus* samples, and compared to DNA of current *Citrullus* species and cultivars *(2 - 44)* (MW 100 bp markers).

PCR Primers

For molecular analysis twelve nSSR (nuclear microsatellite) primer-pairs were used: *Cm*-TC-51 (att ggg gtt tct ttg agg tga / cca tgt cta aaa act cat gtg g); *Cm*-TC-168 (atc att gga tgt ggg att ctc / aca gat gga tga aac ctt agg); *Cm*-ACC-146 (caa cca ccg act act aag t / cga cca aac cca tcc gat aa) (Katzir *et al.* 1996; Danin-Poleg *et al.* 2001). *Bngl*-118-2 (gcc ttc cag ccg caa ccc t / cac tgc atg caa agg caa cca ac); *Bngl*-161 (gct ttc gtc ata cac aca cat tca / atg gag cat gag ctt gca tat tt); *Bngl*-339 (ccaaccgtatcagcatcagc / gcagagctctcatcgtcttctt) (Smith *et al.* 1997). *Phi*-118-2 (atc gga tcg gct gcc gtc aaa / aga cac gac ggt gtg tcc atc); *Phi*-121 (agg aaa atg gag ccg gtg aac ca / ttg gtc tgg acc aag cac ata cac); *Cl*-1-06 (cac cct cct cca gtt gtc att cg / aag gtc agc aaa gcg gca tag g); *Cl*-1-20 (cgcgcgtgaggaccctata / aaccgcctcaatcaattgc);

Cl-2-23 (*gag gcg gag gag ttg aga g / aca aaa caa cga aac cca tag c*) and *Cl*-2-140 (*ctt ttt ctt ctg att tga ctg g / act gtt tat ccc gac ttc act a*) (Jarret *et al*. 1997) (Figures 3 and 4).

ITS sequence analysis was carried out at the ITS1-5.8S-ITS2 loci according to Al-Janabi *et al*. (1994) and aligned to NCBI AJ488232.

For cpDNA analysis two loci of *trna*VAL-*rps*12 (NCBI AJ970307; Al Jabani *et al*. 1994) and *ycf*9-*orf*62 (NCBI AY522531, AY522537 and AY522539; Dane and Liu 2007) were probed.

For flesh color reconstruction *lcyb* gene (*lycopene β-cyclase*) was probed according to Bang *et al*. (2007) for red flesh (NCBI EF183522) and yellow flesh color (NCBI EF183521).

Data Analysis

Cluster analysis was carried out by the SPSS-11 program package using the Average Linkage within group based on the presence versus absence of SSR alleles (Figure 5). Sequence alignments were analyzed by BioEdit Sequence Alignment Editor (NCSU, USA) (Hall *et al*. 1999), MULTALINE (Combet *et al*. 2000) and CLUSTALW (Thompson *et al*. 1994) programs. For SSR repeat search FastPCR Program was used (Kalendar *et al*. 2009). BLAST (*Basic Local Alignment Search Tool*) analysis was carried out by computer program of NCBI (*National Center for Biotechnology Information*, Bethesda, Maryland, USA) (Altschul *et al*. 1990). For dendrogram analysis SPSS (*Statistical Package for the Social Sciences*; Stanford Univ, USA), and MEGA 4 (*Molecular Evolutionary Genetics Analysis*) computer programs were used with Phylogeny NJ (Neighbor Joining) algorithm, and radial and rectangle type of dendrograms were edited (Tamura *et al*. 2007).

CONCLUSION

Genus *Citrullus*

The monotypic genus *Citrullus* is comprised of only four diploid (2n = 4x = 22; 4.25 - 4.54 x 10^8 bp; 0.42 pg DNS) species, including the annual watermelon (*Citrullus lanatus*), the perennial colocynth (*syn.*: bitter apple) (*Citrullus colocynthis*), and two wild species growing in Kalahari Desert, Africa, as the *Citrullus ecirrhosus* with bitter-tasting fruit, and the annual *Citrullus rehmii* with pink and olive green spotted, mandarin sized, non-edible fruits (Robinson and Decker-Walters 1997; Dane and Liu 2007; Feher 1993).

Unlike genus *Citrullus*, species watermelon (*Citrullus lanatus*) comprises diverse varieties, subspecies, mutants and feral forms such as the cultivated watermelon (*C. lanatus lanatus*) (syn.: *C. vulgaris*) with its ancient form of citron melon (*syn.*: African tsamma) (*Citrullus lanatus citroides*), and the Mediterranean seed mutant egusi type watermelon (*C. lanatus mucospermum*) (Kanda 1951; Gusmini *et al*. 2004). Watermelon, citron and colocynth have a history of production in Europe (Wasylikowa and Veen 2004; Creamer 2005; Tóth *et al*. 2007, 2008a,b).

The oldest plant remains with proven human activity have revealed only cereal seeds of wild barley (*H. spontaneum*) and wild emmer (*Triticum dicoccoides*) from 19-23,000 BP

(*Before Present*) at Ohalo II., river Jordan (Nadel *et al.* 1994, 2006; Piperno *et al.* 2004; Weiss *et al.* this volume). The 17,310 ± 310 BP site in Korea (Chungbuk National University, South Korea) revealed the first ancient rice (*Oryza sativa*) seed remains with an extractable amount of aDNA (Suh *et al.* 2000).

The first *Cucurbit* seeds were excavated from the Spirit Cave (Hoabinh, Thailand) including cucumber type *Cucumis* seeds at least 9180 ± 360 BP as analyzed by C^{14} of bamboo charcoal (Gorman 1969).

Ancient Watermelon Seeds Excavated

The primary gene centre for watermelon is not known. However the excavations of five and six thousand year-old seeds of *C. lanatus* in Egypt and Libya imply that domestication might have occurred in Northern Africa (Dane and Liu 2007).

The oldest, 6000-yr-old watermelon (*Citrullus l. lanatus*) seeds were excavated in Helwan (Egypt, Africa), at a site 4000 BCE (Barakat 1990). About 5000-yr-old seed remains were excavated in Uan Muhuggiag, Lybia, Africa from a site 3000 BCE (Wasylikowa and Veen 2004). The most ancient image of watermelon from Pharaohs tomb is known form 3100 - 2100 BCE (Old Kingdom), which shows not round but elongated type melon with green strips (Manniche 1989; Janick *et al.* 2007).

Several watermelon seeds were found in Pharaoh's tombs in Thebes (New Kingdom: 1550-1070 BCE) (stored in Agricultural Museum, Dokki, Giza, Egypt (Warid 1995), and in the pyramid of Tutankhamum ca 1330 BCE (Hepper 1990; Kroll 2000; Vartavan and Amorós 1997).

Hieroglyph of watermelon is known also from 1550 BCE (Warid 1995). The first figures of colocynth (*C. colocynthis*) (known in Arabic as *handel*) were carved into the cedar wood in Solomon's temple (960 - 580 BCE), which is the only poisonous / medical plant displayed in the temple (1 Kings 6:18a).

The Greeks and Romans traveling to Egypt must to have known of watermelons probably without discriminating it from colocynth and citron melon. Pliny II. wrote about a wild (probably the current colocynth) and two types (one with pale green, and the other with grass green rind) of cultivated colocynth (probably the current watermelon), as it has been written: „...*Another kind of wild gourd is called Colocynthis. The fruit is smaller than the cultivated one, and full of seeds. The pale variety is more useful than the grass-green one...*" (Pliny 23-79) (Blake 1981).

The Codex *De Materia Medica* (produced not too long after the time of Pliny) provides nearly 400 color paintings of different plants but no watermelon illustration, only a precise color painting of colocynth which looks very much like the current forms of colocynth (*C. colocynthis*) (Dioscorides 1st CENT and the second 'edition' with color paintings from 512 CE).

Six hundred years later, when the Iberian Peninsula was conquered by the Berbers (Moors) led by Tarik Ibn Ziyad in 711 CE new watermelon types might have entered Europe as recorded in the ancient record of *Book of Agriculture* (Al-Awwam 1158). In this book, two cultivated forms were compared, a black seed type (with dark-green rind which turns black when it ripens) and a red seed type (with green rind which turns to yellow) (Blake 1981).

Watermelon might have also been introduced to Europe through Crusaders (Fischer 1929) led by either Richard I. The Lion-Hearted (the 3rd Crusaders, 1190-1199), or the Hungarian King, Endre II. of the Árpád Dynasty (led the last, 6th Crusaders, 1228-1229). Watermelon spread through Europe quickly and became a very popular and commonly cultivated fruit of the Renaissance Europe, with the first illustration on the frescos in the *Villa Farnesina*, Rome, Italy, 1515-1518 (painted by Giovanni Martini da Udine) (Janick and Paris 2006). It shows watermelons with pale green rind, which indicate an ancient rind type, as a QTL locus (*gs*) responsible for dark-green rind was found to be dominant in the cultivated lines over the light-green rind of the wild watermelons (Hashizume *et al.* 2003).

By 800 CE, watermelons had been introduced to India and by 1100 CE to China. It reached the New World after Columbus' second voyage in 1493 and dispersed quickly among American natives (Blake 1981). One of the most ancient forms of small, round fruit with thin, green rind, red flesh and small black seeds has survived up to the recent times (Gilmore 1919).

One of the oldest watermelon herbarium samples is available from G Bauhin's (1560-1624) collections (about a hundred years earlier than Linnaeus's), who named it *Anguira citrullus* (personal communication, Mark Spencer, The Natural History Museum, London, UK). No watermelon herbarium sample remained from C Linnaeus (1753) collections, who named watermelon as *Cucurbita citrullus*, and colocynth as *Cucumis colocynthis* (personal communication, Arne Anderberg, The Linnean Herbarium, Swedish Museum of Natural History, Stockholm, S).

SSRs of aDNAs

Watermelon seeds excavated at both medieval sites analyzed in this Chapter appeared to be extremely well preserved due to anaerobic conditions in the slime of a deep well covered by water, apparently used as dust holes in the Middle Ages (Gyulai *et al.* 2008) (Figure 1). The herbarium sample seeds from the 19th CENT was stored under precise conditions in glass containers (Vörös 1971). The quality and quantity of aDNA samples showed good and amplifiable DNA ratios (Figure 2).

Twelve microsatellite probes were used for morphological reconstruction of the ancient watermelons (Figure 5 and 9). Allelic diversity of microsatellites were also reliably detected in aDNAs of ancient seagrass (*Posidonia oceanica*) (Raniello and Procaccini 2002). SSRs were also used to morphologically reconstruct 600-yr-old melon (*Cucumis sativus*) (Szabó *et al.* 2005) and millet (*Panicum miliaceum*) (Lágler *et al.* 2005; Gyulai *et al.* this volume). SSR analysis was also applied to herbarium samples of common reed (*Phargmites australis*), about 100-yr-old, to track plant invasion in North America (Saltonstall 2003).

Multiple Sequence Alignment (MSA) computed by MULTALIN (Combet *et al.* 2000) of the (CT)n SSR fragments at the *Cl*1-20 locus (Jarret *et al.* 1997) showed two sharp molecular events occurred during the domestication of watermelon. A (CT)$_3$ deletion in the core sequences of (CT)n resulted in two alleles (*Ia.* and *Ib.*) (Figure 4). Both medieval watermelons and current cultivars # 6, 8, 11 and 30 were heterozygous, and all the rest of the current watermelons were homozygous at this locus.

A part of dinucleotide sequence $(CT)_5$ of the Cl-1-20 allele showed an inversion in all watermelon cultivars that formed a compound SSR $(CT)_n$-C-$(CT)_n$-T-$(CT)_n$ from a putative ancestral SSR $(CT)_{27}$ (Figure 4). These results indicate a step of the endogenous way of elimination of the SSR through fragmentation (Messier *et al.* 1996; Taylor *et al.* 1999).

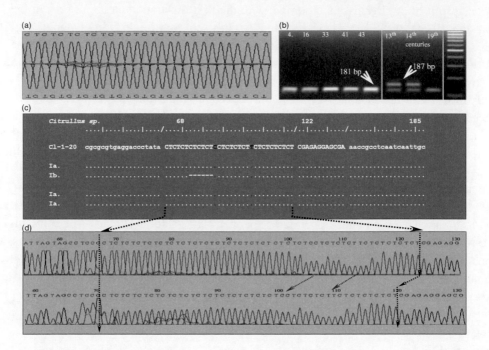

Figure 4. Sequence analysis of aDNAs at the $(CT)n$ SSR locus (*Cl*-1-20). *(a)* Double stranded fluorogram of a $(CT)n$ SSR sequence. *(b)* Samples of agarose gel electrophoresis (0.8%) of *Cl*-1-20 fragments in current *Citrullus* varieties *(# 4-43)*, and compared to archaeological samples of 13th-, 15th- and 19th CENTs (Mw 100 bp ladder from 100 bp). *(c)* Consensus sequence alignments (*indicated by dost*) of *Cl*-1-20 SSR alleles (primer sequences are in italics). A $(CT)_3$ deletion (*underlined*), indicated by (-) created two alleles *Ia.* (187 bp) and *Ib.* (181 bp) (notice fragments on Figure *b*). Both medieval watermelons (13th and 15th CENTs) and current cultivars (# 6, 8, 11 and 30) were *heterozygous* (Ia:Ib), and all the rest of the current watermelons were *homozygous* (Ia:Ia) at this locus. A $(CT)_5$ inversion (*in frames*) formed a compound SSR $(CT)_n$-C-$(CT)_n$-T-$(CT)_n$ from an putatively ancient $(CT)_{27}$ SSR in all watermelons. *(d)* Sequences and fluorograms of the full length alleles *Ia.* $(CT)_{18}$-C-$(CT)_4$-T-$(CT)_5$ (*up*), and *Ib.* $(CT)_{15}$-C-$(CT)_4$-T-$(CT)_5$ (*bottom*). Arrows indicate the shortening of SSRs; and the CC (*blue*) and TT (*red*) motives.

cpDNA in the seed aDNA

Chloroplast genome (cpDNA) specific primers provides highly sensitive methods for analyzing cpDNA in the total (nuclear and organelle) DNA sample without using the former tedious ultracentrifuge separations (Al-Janabi *et al.* 1994; Demesure *et al.* 1995; Dane *et al.* 2007).

Identification of cpDNA haplotypes revealed domestication event in *Citrullus* based on cpDNA at the loci of *ndh*F, *ycf*6-*psb*M, *ycf*9-*trn*G and *atp*A-*trn*R (Dane and Lang 2004);

*ndh*A, *trn*S-*trnf*M and *trn*S-*trnf*M (Dane *et al.* 2004); *G3pdh* (Dane *et al.* 2007); and *trn*S-*trn*G, *atp*A-*trn*R and *trn*E-*trn*T (Dane and Liu 2007).

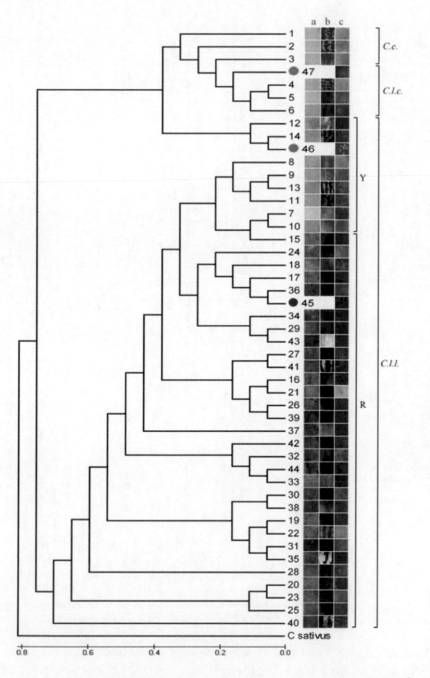

Figure 5. Molecular dendrogram (SPSS-17) based on SSR analysis of aDNA samples of 13th CENT *(45)* and 15th CENT *(46)* watermelons *(Citrullus lanatus lanatus; C.l.l.)*, and 19th CENT citron melon *(Citrullus lanatus citroides; C.l.c.)*. Ancient samples were compared to current *cv.* colocynth *(Citrullus colocynthis; C.c.) (1-3)*, citron melons *(3-6)* and watermelons *(7-44)*. Colors of flesh *(a)*, rind *(b)* seed coat *(c)*, clusters of flesh colors (R – red, Y – yellowish white) and an outgroup *Cucumis sativus* are indicated.

Table 1. SNPs (single nucleotide substitutions) at the sites of $tRNA^{Val}$ **-** $rps12$ **(a-f) and** $ycf9$ - $orf62$ **(g - i) of cpDNAs of** *Citrullus* **samples**

#	Cultivars	Access. #	sp.	a	b	c	d	e	f	g	h	i
		NCBI #AJ970307	*C.s.*	G	G	A	G	A	A			
01	Finn	RCAT036168	*C. c.*	T	T	.	.	.	.	T	+	-
02	Belga	RCAT036172	*C. c.*	T	T	.	.	.	.	T	+	-
03	Portugál	RCAT035547	*C. c.*	T	T	.	.	.	.	T	+	-
47	**19th CENT**		*C. l. c.*	T	T	.	.	.	.	A	-	+
04	Szeged	RCAT036099	*C. l. c.*	T	T	.	.	.	.	A	-	+
05	Bánát	RCAT035235	*C. l. c.*	T	T	.	.	.	.	A	-	+
06	Újszilvás	RCAT055816	*C. l. c.*	T	T	.	.	.	.	A	-	+
07	Bácsbokod	RCAT035917	*C. l. l.*	T	.	.	.	.	.	A	-	+
10	Dévaványa	5101./02	*C. l. l.*	T	.	.	.	.	.	A	-	+
16	Biri	RCAT035114	*C. l. l.*	T	.	.	.	.	.	A	-	+
18	Charleston gray	00263/05	*C. l. l.*	T	.	.	.	.	.	A	-	+
21	Ukrainskij	RCAT036149	*C. l. l.*	T	.	.	.	.	.	A	-	+
24	Háromfa	RCAT034754	*C. l. l.*	T	.	.	.	.	.	A	-	+
25	Debrecen	RCAT035111	*C. l. l.*	T	.	.	.	.	.	A	-	+
26	Sibiriak	RCAT036098	*C. l. l.*	T	.	.	.	.	.	A	-	+
27	Nagyecsed	RCAT034775	*C. l. l.*	T	.	.	.	.	.	A	-	+
34	Kömörö	RCAT034762	*C. l. l.*	T	.	.	.	.	.	A	-	+
39	Gyöngyös	RCAT034969	*C. l. l.*	T	.	.	.	.	.	A	-	+
41	Kibéd	5172./02	*C. l. l.*	T	.	.	.	.	.	A	-	+
44	Korai kincs	00255/05	*C. l. l.*	T	.	.	.	.	.	A	-	+
45	**13th CENT**		*C. l. l.*	T	.	.	A	.	.	A	-	+
46	**15th CENT**		*C. l. l.*	T	.	.	A	.	.	A	-	+
12	Belyj dlinnij	RCAT036152	*C. l. l.*	T	.	.	A	.	.	A	-	+
14	Csárdaszállás	RCAT035113	*C. l. l.*	T	.	.	A	.	.	A	-	+
17	Klondike	RCAT036096	*C. l. l.*	T	.	.	A	.	.	A	-	+
36	Kecskeméti v	00259/05	*C. l. l.*	T	.	.	A	.	.	A	-	+
08	Napsugár	00257/05	*C. l. l.*	T	.	G	.	.	.	A	-	+
15	Tura	RCAT035389	*C. l. l.*	T	.	G	.	.	.	A	-	+
20	Túrkeve	RCAT035112	*C. l. l.*	T	.	G	.	.	.	A	-	+
29	Hevesi	00258/05	*C. l. l.*	T	.	G	.	.	.	A	-	+
23	Marsowszky	00256/05	*C. l. l.*	T	.	G	.	.	.	A	-	+
09	Sándorfalva	RCAT036105	*C. l. l.*	T	.	.	.	G	G	A	-	+
13	Ráckeve	RCAT055812	*C. l. l.*	T	.	.	.	G	G	A	-	+
38	Pusztadobos	RCAT035146	*C. l. l.*	T	.	.	.	G	G	A	-	+
30	Nagyvárad	RCAT034767	*C. l. l.*	T	.	.	.	G	G	A	-	+
32	Orosháza	RCAT035862	*C. l. l.*	T	.	.	.	G	G	A	-	+
11	Szentesi sárga	00260/05	*C. l. l.*	T	.	G	.	G	G	A	-	+
19	Taktaharkány	RCAT034790	*C. l. l.*	T	.	G	.	G	G	A	-	+
22	Szirma	RCAT034782	*C. l. l.*	T	.	G	.	G	G	A	-	+
28	Nagykálló	RCAT034785	*C. l. l.*	T	.	G	.	G	G	A	-	+
31	Nyírbátor	RCAT035155	*C. l. l.*	T	.	G	.	G	G	A	-	+
33	Rákóczifalva	RCAT035145	*C. l. l.*	T	.	G	.	G	G	A	-	+
35	Nyíregyháza	RCAT034778	*C. l. l.*	T	.	G	.	G	G	A	-	+
37	Ilk	RCAT035236	*C. l. l.*	T	.	G	.	G	G	A	-	+
40	Crimson sweet	00262/05	*C. l. l.*	T	.	G	.	G	G	A	-	+
42	Sugar baby	00261/05	*C. l. l.*	T	.	G	.	G	G	A	-	+
43	Lipót	RCAT034970	*C. l. l.*	T	.	G	.	G	G	A	-	+

(a) $102,029^{th}$; *(b)* $102,182^{nd}$; *(c)* $102,191^{st}$; *(d)* $102,193^{rd}$; *(e)* $102,217^{th}$; and *(f)* $102,218^{th}$ nt of total cpDNAs. *(g)* SNP at 320th nt, *(h)* +/- of a 131 nt long sequence, and *(i)* +/- an ATAGC sequence at *ycf9* - *orf62* site. Samples, *(01-03)* colocynth *(C.c., Citrullus colocynthis)*. *(04-06)* citron melon *(C.l.c., C. lanatus citroides)*. *(07-44)* watermelon *(C.l.l., C.l.lanatus)*. *(45)* 13th -, *(46)* 15th -, and *(47)* 19th CENT samples. Accession # of Gene Bank Tápiószele (Hungary) are listed. Outgrup: *Cucumis sativus (C.s.)* (NCBI # AJ970307).

All *Citrullus* sample had a G→T substitution at the $102,029^{th}$ nt compared to gene bank data of *Cucumis sativus* (AJ970307) (Table 1, *a*,). Five new haplotypes based on nucleotide substitution at $tRNA^{Val}$ - *rps12* locus were also identified (Table 1, *b-f*,).

Samples of colocynth (# 1-3), current citron melon (# 4-6) and the 19th CENT citron melon clustered in a separate group as a result of a G→T substitution at $102,182^{th}$ nt (Table 1, *b*). This substitution turned out to be sub-species specific marker between citron melon and watermelon as all watermelon sample had G at this locus. Medieval samples and four current watermelons clustered in separate group due to the G→A substitution at the 102,193 nt, which results might indicate unique date for cytoplasm reconstruction of ancient samples (Figure 6).

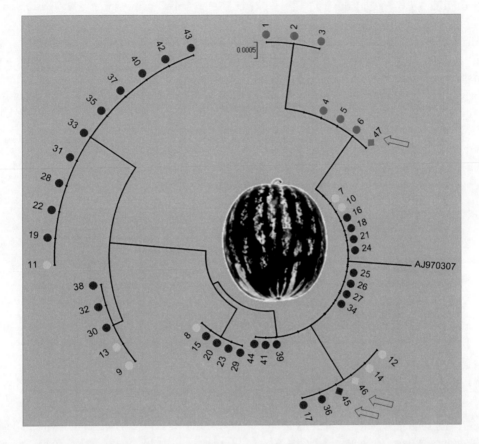

Figure 6. cpDNA. Molecular dendrogram (MEGA 4, Tamura *et al.* 2007) based on SNPs of two loci of cpDNA at *ycf9-orf62*, and *tRNA*^Val - *rps12* of current *Citrullus* species and cultivars *(1 - 44)* (listed in Table 1), and compared to archaeological samples from *(45)* 13th -, *(46)* 15th -, and *(47)* 19th CENTs. (arrows) *(1-3)* colocynth *(Citrullus colocynthis)*, *(4-6)* citron melons *(C. lanatus citroides)*, *(7-44)* current watermelon *(C. l. lanatus)* cultivars. Outgrup: *Cucumis sativus* (NCBI # AJ970307). Flesh colors are indicated by circles based on *lcyb* gene sequences (Figure 8): green (whitish flesh), yellow (yellow flesh), and red (red flesh).

Five current watermelons grouped in one cluster as the results of A→G substitutions at the 102,217 nt and 102,218 nt. Eleven watermelon cultivars clustered in the final group with triple A→G substitutions (102,191; 102,217; and 102,218 nt). (Table 1, Figure 6).

As medieval watermelons did not create a separate group (Fig. 6), but clustered to current cultivars *Belyj dlinnij* (#12), *Csárdaszállás* (#14), *Klondike* (#17) and *Kecskeméti vörös* (#36), it indicates that medieval melon haplotype, as probably the most ancient type in Europe, is still present in current varieties.

ITS of aDNAs

The highly conserved ribosomal rDNA sequences of current cultivars of colocynth (# 1-3), citron melons (# 4-6) and the 19th CENT citron melon showed the same SNP pattern at the ITS1-5.8S-ITS2 locus (Figure 7).

Current citron melon cv. *Bánát* (#5) was found to carry a watermelon-specific rDNA alleles at heterozygote form, which might indicate an evolutionary step from bitter tasted citrons towards watermelons.

In contrast, 15th cent watermelon, and current watermelon cv. Túrkeve (#20) was found to carry a citron-specific rDNA allele at heterozygote form, which indicates that red flesh watermelon cv. Túrkeve (#20) is the most ancient watermelon type among the accessions studied.

The 15th CENT watermelon also showed heterozygote alleles at this locus. 13th CENT watermelon was homozygous with less SNPs (Figure 7).

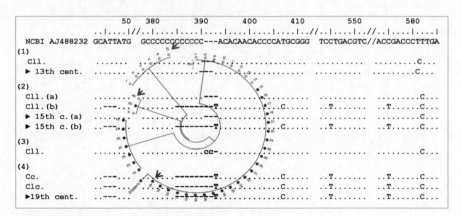

Figure 7. ITS sequence analysis and ITS dendrogram (MEGA 4, Tamura *et al.* 2007) with indications of flesh colors (*Figure 9*) (outgroup NCBI AJ488232) of ancient ▶13th *(#45)*, ▶15th *(#46)* and ▶19th *(#47)* CENTs and current *Citrullus* samples. Four allele combinations (*1 - 4*) were detected at the ITS1-5.8S-ITS2 locus of rDNA. Cll. - *Citrullus l. lanatus (7-44)* . Cc. - *Citrullus colocynthis (1-3)*; Clc. - *Citrullus lanatus cytroides (4-6)*.

What Type of Melons Did King Mathias (1443 FEB 23 – 1490 APR 06) Eat in the Middle Ages?

The enzyme LCYB (*lycopene β-cyclase*) encoded by *lcyb* gene play a central role in plant color development by converting lycopene to carotenoids with ring structure. SNP markers in *lcyb* gene at the 1182th base pair (bp) were found to be A=T (Adenine / Thymine) in all CY (canary yellow fleshed) watermelons with both homozygous dominant (*lcyB/lcyB*) and heterozygous (*lcyB/lcyb*) alleles. However, all red flesh watermelons with homozygous (*lcyb/lcyb*) alleles had G≡C (Guanine / Cytosine) bp at this locus (Bang *et al.* 2007) resulting in an amino acid substitution from *Phe* to *Val* at the 226th amino acid locus. Another SNP at the 518th bp with the opposite way of substitutions (G≡C in the canary yellow fleshed melons, and T=A in the red ones) were also found to be identical (Bang *et al.* 2007).

SNPs of *lcyb* gene of watermelons studied in this Chapter revealed that all current watermelons with red flesh (# 15 - 44) and the 13th CENT sample had recessive homozygote alleles (*lcyb/lcyb*) at both SNP loci with T=A at the 518th bp and G≡C at the 1182th bp, respectively (Figure 8). The 15th CENT and 19th CENT samples along with all current colocynths (#1-3), citrons (#4-6), and yellow flesh watermelons (#7-14) had homozygous dominant CY-alleles at both SNP loci of *lcyb* gene with G≡C bp at the 518th bp, and T=A at 1182th bp, respectively (Figure 8).

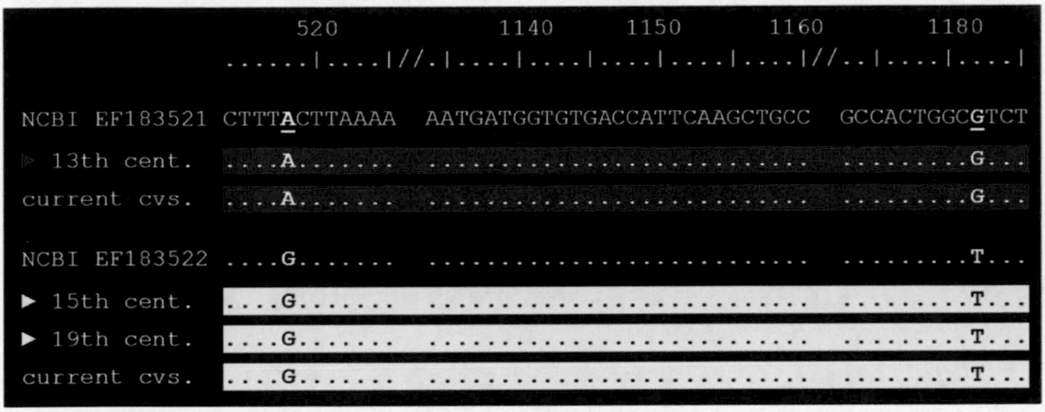

Figure 8. Flesh color reconstructions from aDNAs of ►13th-, ►15th- and ►19th CENT *Citrullus* seeds based on SNPs of the *lcy*b gene at the 518th and 1182th bps.

Flesh colors of watermelons varies from white to yellow, canary yellow, salmon yellow, orange mainly due to pigment of xanthophylls. Pink, red and purple colors are mainly due to pigments of lycopenes. Genes coding for white flesh color (*w*) were QTL-mapped on chromosome 6 (*syn*.: linkage group 6) (Hashizume *et al.* 1996). Genes responsible for yellow and red color were mapped on chromosome 2, which might be the reason of transition forms of yellow, canary yellow, pale yellow (Hashizume *et al.* 2003). QTL responsible for red flesh color had another locus on chromosome 8. This locus showed genetic linkage with QTL locus for high sugar (brix value) content (Hashizume *et al.* 2003). Red fleshed watermelon appeared in the painting of *Still Life with Melons and Carafe of White Wine* (1603) painted by Caravaggio (Janick 2004; Janick *et al.* 2007). This result might indicate the reason of dominance of red flesh color watermelon cultivars over white and yellow colored ones, as selection for sweeter watermelons during domestication has been coupled with selection for red flesh color at the same time (Hashizume *et al.* 2003). Some further genetic loci for color determination were recently determined by breeding tools (crossings), namely Y (red, dominant), y° (orange, recessive), y (salmon yellow, recessive), C (canary yellow, dominant) and c (red, recessive), respectively (reviewed in Bang *et al.* 2007).

To conclude, 701 DNA fragments of 23 alleles at twelve SSR loci, and sequence analyses of cpDNA, RAPD, ITS and *lcyb* gene revealed that aDNA of 13th CENT watermelon (Debrecen) showed close molecular similarity to current red fleshed watermelons; the 15th CENT sample (Budapest) showed close similarity to current yellow fleshed watermelons (Figure 9). These results reflect the preferential cultivation of both red-flesh and yellow-flesh

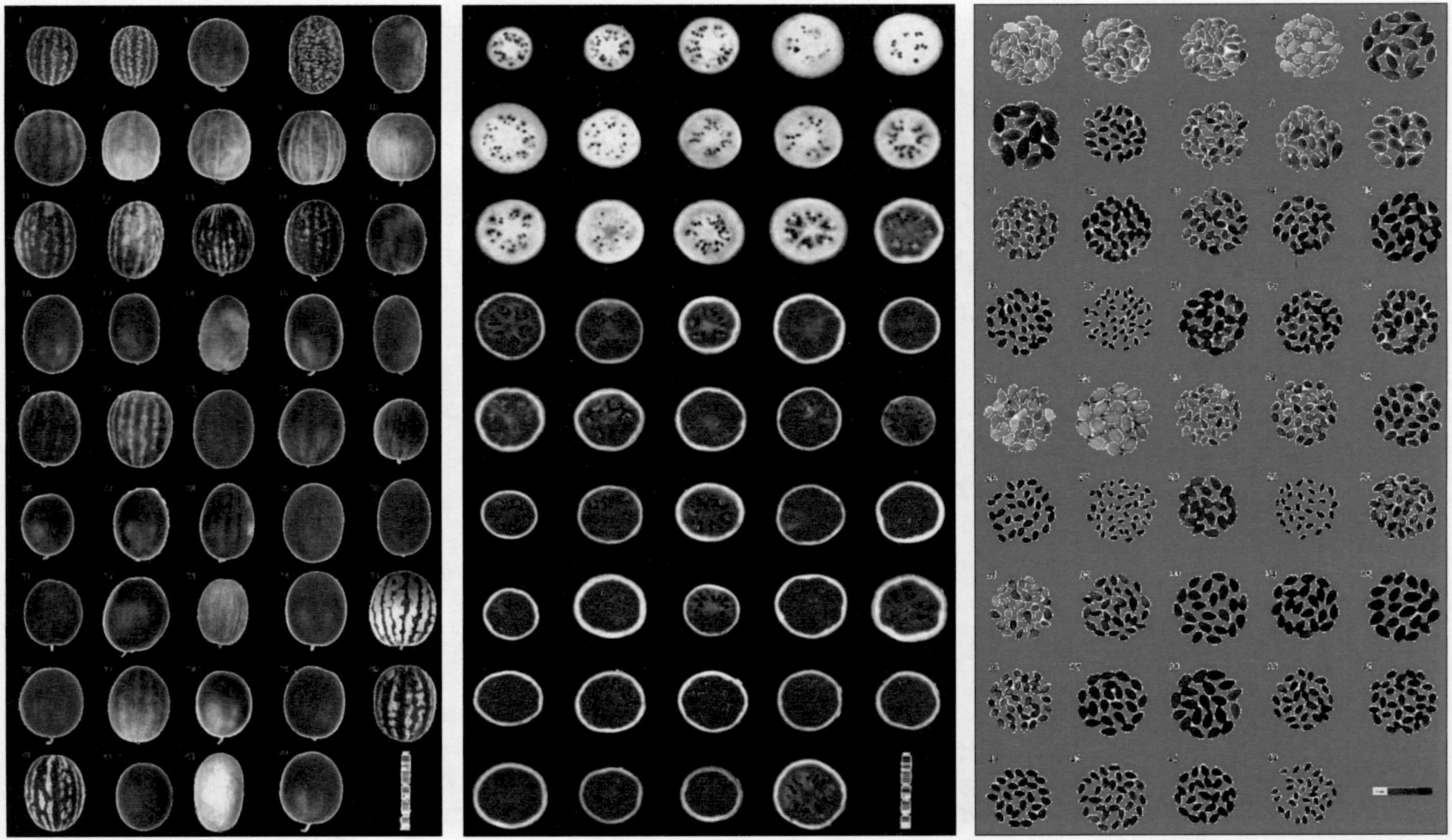

Figure 9. Rind (*left*) flesh (*middle*) and seed (*right*) types of current *Citrullus* species and cultivars (listed in Table 1) at ripening time used for comparative analyses (color bars 25 cm).

watermelons in the Middle Ages of Hungary. The 170-yr-old herbarium sample (Pannonhalma) showed close molecular similarity to citron melons (*Citrullus l. citroides*), which reflects the importance of citron melon as fodder in the Middle-Ages in Hungary and Europe.

ACKNOWLEDGMENTS

The authors wish to thank Dr. Zoltán Szabó for his contribution and comments on the manuscript.

REFERENCES

Al-Janabi SM, Honeycutt RJ, Peterson C, Sobral BWS (1994) Phylogenetic analysis of organellar DNA sequences in the Andropogoneae: *Saccharum. Theor Appl Genet.* 88: 933-944.

Al-Awwam Ibn (1158) Le Livre de l'agriculture, translated by Clement-Mullet; 2 tomes in 3 Vols., Paris (1864-1867). Reprints (1802, 1988). Libro de agricultura / su autor el doctor excelente Abu Zacaria Iahia; [traducido al castellano y anotado por Josef Antonio Banqueri ; estudio preliminar y notas , J. E. Hernández Bermejo y E . García Sánchez]. Madrid.

Altschul SF, Gish W, Miller W, Myers EW, Lipman DJ (1990) Basic local alignment search tool. *Jourmnal of Molecular Biology* 215(3): 403-410.

Bang H, Kim S, Leskovar D, King S (2007) Development of a codominant CAPS marker for allelic selection between canary yellow and red watermelon based on SNP in lycopene B-cyclase gene. *Mol Breeding* 20: 63-72.

Barakat H (1990) Appendix IV – plant remains from El-Omari, in: F Debono, B Mortensen (Eds.), El-Omari. *A Neolithic Settlement and Other Sites in the Vicinity of Wadi Hof, Helwan.* Verlag Philipp von Zabern ISBN 3-8053-1119-2, Mainz am Rhein, 1990, pp. 109-114.

Biss P, Freeland J, Silvertown J, McConway K, Lutman P (2003) Successful amplification of rice chloroplast microsatellites from century-old grass samples from the park grass experiment. *Plant Mol Biol Rep* 21: 249-257.

Bisztray GyD, R Bacsó, P Bodor, G Facsar, F Gyulai, I Velich (2004) Archaeobotanical and genetical methods to analyse 600-years-old seeds of horticultural plants. Book of Abstracts 5th IVCHB Symposium, *In Vitro Culture and Horticultural Breeding*, 12-17. September 2004, Debrecen, Hungary, p. 212.

Blake LW (1981) Early Acceptance of Watermelon by Indians of the United States. *Ethnobiology* 1: 193-199.

Brown TA (1999) How ancient DNA may help in understanding the origin and spread of agriculture. *Proceedings of the Royal Society of London, Series B* 354: 89-98.

Combet C, Blanchet C, Geourjon C, Deléage G (2000) NPS@: Network Protein Sequence Analysis. *Trends in Biochem Sci* 25(3): 147-150.

Cooper A, Poinar HN (2000) Ancient DNA: Do it right or not at all. *Science* 289: 1139.

Cooper A, Lalueza-Fox C, Anderson S, Rambaut A, Austin J, Ward R (2001) Complete mitochondrial genome sequences of two extinct moas clarify ratite evolution. *Nature* 409: 704-707.

Creamer J (2005) *Searching for a better melon* (by G Gyulai and F Dane). AAES Impact, Augustus, Auburn, USA, p.2.

Dane F, Lang P (2004) Sequence variation at cpDNA regions of watermelon and related species: implications for the evolution of *Citrullus* haplotypes. *Amer J Bot* 91: 1922-1929.

Dane F, Lang P, Backtiyarova R (2004) Comparative analysis of chloroplast variability in wild and cultivated *Citrullus* species. *Theor Appl Genet* 108: 958-966.

Dane F, Liu J (2007) Diversity and origin of cultivated and citron type watermelon (*Citrullus lanatus*). *Genet Resour Crop Evol* 54: 1255-1265.

Dane F, Liu J, Zhang C (2007) Phylogeography of the bitter apple, *Citrullus colocynthis*. *Genet Resour Crop Evol* 54: 327-336.

Danin-Poleg Y, Reis N, Tzuri G, Katzir N (2001) Development and characterization of microsatellite markers in *Cucumis*. *Theor Appl Genet* 1002: 61-72.

Demesure B, Sozi N, Petit RJ (1995) A set of universal primers for amplification of polymorphic non-coding regions of mitochondrial and chloroplast DNA in plants. *Mol Ecol.* 4: 129-131.

Dioscorides (512) (*true name Pedianos Dioskurides*) De Materia Medica (with nearly 400 color paintings) (1st century, Painting from the year 512). Der Wiener Dioskurides: Codex medicus Graecus 1 der Österreichischen National bibliothek (1998-1999) commentary by Otto Mazal (published in a reduced two-volume facsimile by Akademische Druck-und Verlagsanstalt).

Don RH, Cox PT, Wainwright BJ, Baker K, Mattick JS (1991) Touchdown PCR to circumvent spurious priming during gene amplification. *Nucleic Acids Res* 19: 4008.

Erlich HA, Gelfand D, Sninsky JJ (1991) Recent advances in the polymerase chain reaction. *Science* 252: 1643-1651.

Feher T (1993) Watermelon. In: G Kalloo and BO Bergh (Eds.), Genetic Improvement of Vegetable Crops, pp. 295-315. Pergamon Press Ltd., Tarrytown, New York.

Fischer H (1929) Mittelalterliche Pflanzenkunde. München: Verlag der Münchner Drucke.

Gilmore MR (1919) Uses of plants by the Indians of the Missouri river region. Univ. of Nebraska Press, Lincoln. (reprinted from the 33rd Annu. Report Bur. Amer. Rthn., Washington, 1977).

Gorman CF (1969) Hoabinhian: a people-toll complex with early plant associations in southeast Asia. *Science* 163: 671-673.

Gugerli F, Parducci L, Petit RJ (2005) Ancient plant DNA: review and Prospects. *New Phytologist* 166: 409-418.

Gusmini G, Wehner TC, Jarret RL (2004) Inheritance of Egusi Seed Type in Watermelon. *J Hered* 95: 268-270.

Gyulai G, Magda A, Kiss J, Gyulai G, Holly L, Heszky L (2001) DNS izolálás és PCR-amplifikáció 700 éves növény magvakból. VII. Növénynemesítési Tudományos Napok, Budapest, p.89.

Gyulai G, Waters L, Dane F (2008) Ancient *Citrullus* DNA-unlocking domestication events. Fulbright Grant AY 2005-2006. Fulbright Year Book 2008, Budapest, pp 25-42.

Hall TA (1999) BioEdit: a user-friendly biological sequence alignment editor and analysis program for Windows 95/98/NT. *Nucleic Acids Symposium Ser* 41: 95-98.

Hartyányi BP, Nováki Gy (1975) Samen- und fruchtfunde in Ungarn von der neusteeinzeit bis zum 18. jahrhundert. *Agrártört Szemle, Budapest* 17: 1-88.

Hashizume T, Shimamoto I, Harusima, Yui M, Sato T, Imai T, Hirai M (1996) Construction of a linkage map for watermelon (*Citrullus lanatus*) using random amplified polymorphic DNA (RAPD). *Euphytica* 90: 265-273.

Hashizume T, Shimamoto I, Hirai M (2003) Construction of a linkage map and QTL analysis of horticultural traits for watermelon [*Citrullus lanatus* (Thunb.) Matsum & Nakai] using RAPD, RFLP and ISSR markers. *Theor Appl Genet* 106: 779-785.

Hepper FN (1990) Pharaoh's flowers. The botanical treasures of Tutankhamun. Royal Botanic Gardens, Kew, London. 80 pp. ISBN 0 11 250040 4.

Horváth L, Gyulai G, Szabó Z, Lágler R, Tóth Z, Heszky L (2007) Morfológiai diverzitás sárgadinnyében (*Cucumis melo*); egy középkori típus fajtarekonstrukciója. *Agrártud Közl* 27: 84-90.

Janick J (2004) Caravaggio's fruit. A mirror on Baroque horticulture. *Chronica Horticulturae* 44: 9-15.

Janick J, Paris HS (2006) The Cucurbit Images (1515 - 1518) of the Villa Farnesina, Rome. *Annals of Botany* 97: 165-176.

Janick J, Paris HS, Parrish DC (2007) The Cucurbits of Mediterranean Antiquity: Identification of Taxa from Ancient Images and Descriptions. *Annals of Botany*: 100: 1441-1457.

Jarret RL, Merrick LC, Holms T, Evans J, Aradhya MK (1997) Simple sequence repeats in watermelon (*Citrullus lanatus* (Thunb.) Matsum. & Nakai). *Genome* 43: 433-441.

Kalendar R, Lee D, Schulman AH (2009) FastPCR Software for PCR Primer and Probe Design and Repeat Search. *Genes, Genomes and Genomics* 3(1): 1-14. [http://primerdigital.com/fastpcr.html]

Kanda T (1951) The inheritance of seed-coat colouring in the watermelon. *Jpn J Genet* 7: 30-48.

Katzir N, Danin-Poleg Y, Tzuri G, Karchi Z, Lavi U, Cregan PB (1996) Length polymorphism and homologies of microsatellites in several *Cucurbitaceae* species. *Theor Appl Genet* 93: 1282-1290.

Kroll H (2000) Literature on archaeological remains of cultivated plants (1998/1999) *Veget Hist Archaeobot* 9: 31-68.

Lágler R, Gyulai G, Humphreys M, Szabó Z, Horváth L, Bittsánszky A, Kiss J, Holly L, Heszky L (2005) Morphological and molecular analysis of common millet (*P. miliaceum*) cultivars compared to an aDNA sample from the 15th century (Hungary). *Euphytica* 146: 77-85.

Linnaeus C (1753) Species Plantarum. 2 vols. Salvius, Stockholm. Facsimile edition (1957 - 1959), Ray Society, London.

Mannice L (1989) An ancient Egyptian herbal. Univ. Texas Press, Austin U.S.A.

Messier W, Li S-H, Stewart C-B (1996) The birth of microsatellites. *Nature* 381: 483.

Meszter ZR (2006) Phenotype reconstruction of watermelon using bioinformatics tools. MSc Thesis, pp66, Cranfield University, UK. Supervisors C Bessant and G Gyulai.

Nadel D, Danin A, Werker E, Schick T, Kislev ME, Stewart K (1994) 19,000-Year-Old Twisted Fibers From Ohalo II. *Current Anthropology* 35: 451-458.

Nadel D, Grinberg U, Boaretto E, Werker E (2006) Wooden objects from Ohalo II (23,000 cal BP), Jordan Valley, Israel. *J Human Evol* 50: 644-662.

Piperno DR, Weiss E, Holst I, Nadel D (2004) Processing of wild cereal grains in the Upper Palaeolithic revealed by starch grain analysis. *Nature* 430: 670-673.

Pliny II (*Plinius Gaius Secundus*) (23-79) Historia Naturalis. The Historie of The World. Vol. 1 -10; Book I-XXXVII. (in Latin, with English translation by WHS Jones, 1939-1963). Loeb Classical Library, William Heinemann Ltd., London, UK.

Raniello R, Procaccini G (2002) Ancient DNA in the seagrass *Posidonia oceanica*. *Marine Ecology – Progress Series* 227: 269-273.

Robinson RW, Decker-Walters DS (1997) Cucurbits. CAB International, pp.240, Cambridge. ISBN-10: 0851991335.

Saltonstall K (2003) Microsatellite variation within and among North American lineages of *Phragmites australis*. *Mol Ecol* 12: 1689-1702.

Schermann Sz (1966) Magismeret (*Seed morphology*), Vol. I, II. Akadémiai Kiadó, Budapest.

Schlumbaum A, Tensen M, Jaenicke-Despres V (2007) Ancient plant DNA in archaeobotany. *Veget Hist Archaeobot* 17: 233-244.

Smith JSC, Chin ECL, Shu H, Smith OS, Wall SJ, Senior ML, Mitchell SE, Kresowitch S, Ziegle J (1997) An evaluation of the utility of SSR éoci as molecular markers in maize (*Zea mays* L.): comparison with data from RFLPs and pedigree. *Theor Appl Genet* 95: 163-173.

Suh HS, JH Cho, YJ Lee, MH Heu (2000) RAPD variation of 13,010 and 17,310 year-old carbonized rice. 4th International Rice Genetics Symposium, Manilla, Philipines, Oct. 22-27.

Szabó Z, Gyulai G, Humphreys M, Horváth L, Bittsánszky A, Lágler R, Heszky L (2005) Genetic variation of melon (*C. melo*) compared to an extinct landrace from the Middle Ages (Hungary) I. rDNA, SSR and SNP analysis of 47 cultivars. *Euphytica* 146: 87-94.

Tamura K, Dudley J, Nei M, Kumar S (2007) MEGA4: Molecular Evolutionary Genetics Analysis (MEGA) software version 4.0. *Mol Biol Evol* 24: 1596-1599.

Taylor JS, Durkin JMH, Breden F (1999) The Death of a Microsatellite: A Phylogenetic Perspective on Microsatellite Interruptions. *Mol Biol Evol* 16: 567-572.

Thompson JD, Higgins DG, Gibson TJ (1994) CLUSTAL W: improving the sensitivity of progressive multiple sequence alignment through sequence weighting, positions-specific gap penalties and weight matrix choice. *Nucleic Acids Research* 22: 4673-4680.

Threadgold J, Brown TE (2003) Degradation of DNA in artificially charred wheat seeds. *J Archaeol Sci* 30: 1067-1076.

Tóth Z, Gyulai G, Szabó Z, Horváth L, Heszky L (2007) Watermelon (*Citrullus l. lanatus)* production in Hungary from the Middle Ages (13th century). *Hungarian Agricultural Research* 4: 14-19.

Tóth Z, Gyulai G, Szabó Z, Gyulai F, Heszky L (2008a) New *Citrullus* haplotypes at the *tRNA*-Val – *rps*12 locus of cpDNA. In: Cucurbitaceae 2008, Proceedings of the IXth EUCARPIA meeting on genetics and breeding of Cucurbitaceae (Pitrat M, ed), INRA, Avignon (France), May 21-24th, pp.335-340.

Tóth Z, Gyulai G, Szabó Z, Bittsánszky A, Heszky L (2008b) Genotype (nSSR) and haplotype (cpDNA) identification in watermelons (*Citrullus l. lanatus*). General Meeting EUCARPIA, Valencia, Spain, pp. 253-257.

Vartavan C de, Amorós AV (1997) Codex of ancient Egyptian plant remains. Codex des restes végétaux de l'Egypte ancienne. London, 401 pp.

Vörös L (1971) Seed collection of Pannonhalma High School from the 1830's (in Hungarian). *Bot. Közl.* Budapest, Hungary 58: 179-180.

Wasylikowa K, van der Veen M (2004) An archaeobotanical contribution to the history of watermelon, *Citrullus lanatus* (Thunb.) Matsum. & Nakai (syn. C. vulgaris Schrad.). *Veget Hist Archaeobot* 13: 213-217.

Warid WA (1995) Vegetable species known to the ancient Egyptians. *Acta Hort* 391: 273-290.

Yang H (1997) Ancient DNA from Pleistocene fossils: preservation, recovery, and utility of ancient genetic information for quaternary research. *Quaternary Science Reviews* 16: 1145-1161.

In: Plant Archaeogenetics
Editor: Gábor Gyulai, pp. 89-104

ISBN: 978-1-61122-644-7
© 2011 Nova Science Publishers, Inc.

Chapter 8

GENETIC AND MORPHOGENETIC RECONSTRUCTION OF 15TH CENT MELON (*Cucumis melo*) FROM SEED aDNA

Gábor Gyulai[1,*], Zoltán Szabó[1,2] and Ottó Törjék[1,3]

[1]St. Stephanus University, Institute of Genetics and Biotechnology,
Gödöllő 2103, Hungary

[2]Agricultural Biotechnology Center, Szent-Györgyi Albert 4,
Gödöllő 2100, Hungary

[3]Institute of Plant Biology, Biological Research Center, Szeged 6726, Hungary

ABSTRACT

ITS, SSR, RAPD and sequence analyses of *a*ncient melon DNA (aDNA) extracted from 600-yr-old seed remains recovered from the 15th CENT site in Budapest (Hungary) were analyzed. An aseptic incubation of seeds followed by ITS (*i*nternal *t*ranscribed *s*pacer) analysis was used to exclude the exogenously and endogenously contaminated seeds and to detect SNPs (*s*ingle *n*ucleotide *p*olymorphism) in ITS1-5.8S-ITS2 region (667 bp) of rDNA (*r*ibosomal DNA). SNPs of ITS1 were observed at the 94-95th bp (GC to RC, AC, AG, AS, GG, GS, RS); and of ITS2 at 414th bp (A-to-T substitution), 470th bp (T to Y or C), 610th bp (A to R or G) and 633rd bp (A-to-G transition). A medieval sample was compared to 47 current melon cultivars and landraces with a final aim of genotype and phenotype reconstruction of the medieval melon. For RAPD analysis, 60 primers from primer sets OP-A, -K and -M, (1 to 20 in each case) were applied resulting in a total of 105 fragments in the 15th CENT and current melons. Of them, 61 were monomorphic and 44 showed polymorphisms. Microsatellites at nuclear *s*imple *s*equence *r*epeats (SSR) were detected by *a*utomated *l*aser *f*luorometer (ALF). Eight of the twenty SSR primer pairs tested have amplified 40 microsatellite alleles with a total of 463 fragments in the medieval and current melons. The number of alleles per SSR loci ranged from 2 to 7 with

---

* E-mail: gyulai.gabor@mkk.szie.hu

an average of 5.7 including *Cm*CT44 (2 alleles), *Cm*AG59 (5 alleles), *Cm*GA104 (5 alleles), *Cm*CT134 (4 alleles), *Cm*TA134 (6 alleles), *Cm*CTT144 (7 alleles), *Cm*TC168 (6 alleles) and *Cm*CT170 (5 alleles). Sequence analysis of the SSR alleles at the dinucleotide (CT)n and trinucleotide (CTT)n loci showed different fragment lengths depending on changes in the number of core unit. The length of SSRs did not show a time dependent variation in lengths as 15th CENT melon showed SSRs of intermediate sizes compared to current cultivars. Molecular dendrogram, based on the presence versus absence of SSR alleles, revealed that medieval melon had the closest genetic similarity to a current melon cultivar '*Hógolyó*' (#24) of *inodorus* fruit type (also called winter melon, fist-size melon) with smooth yellow rind, and green flesh color. The results indicate the importance of winter melons in the medieval Europe. As the cultivation of melon in Europe started only in the 13th CENT, however, the Hungarians already named melon (*dinnye*) in the early 11th CENT, the melon samples recovered from the 15th CENT Budapest may originate from one of the oldest cultivated melons in Europe.

Keywords: *A*ncient DNA (aDNA), genotype, ITS (*i*nternal *t*ranscribed *s*pacer in ITS1-5.8S-ITS2 region of rDNA), microsatellites (SSR) – *s*imple *s*equence *r*epeat), phenotype, RAPD (*r*andom *a*mplified *p*olymorphic *D*NA), rDNA (*r*ibosomal DNA), SNP (*s*ingle *n*ucleotide *p*olymorphism)

INTRODUCTION

Archaeological analysis of melon (*Cucumis melo* L, 2n = 4x = 24) has been reported from different excavation sites. The first *Cucurbit* seeds were excavated from the Spirit Cave (Hoabinh, Thailand), including cucumber type *Cucumis* seeds at least 9180 ± 360 BP (*Before Present*), as analyzed by C^{14} of bamboo charcoal (Gorman 1969).

Melon seeds were excavated from 3000 BCE in Iran (Constantini 1977, in Kroll 1999, 2000). Early records of Egyptian mural paintings and Chinese writings appeared from 2000 BCE (Keimer 1924). Greeks and Romans did not grow melons but imported them from Persia or the Caucasus (Stepansky *et al.* 1999), nevertheless *Cucumis* (melon or cucumber) seed remains were recovered in Tiryns (Greece) 1300-1200 BCE (Kroll 1982). From the late Roman age, 3rd CENT CE melon seeds were excavated in Budapest-*Aquincum* (Szabó *et al.* 2004, 2005).

Domestication of melon started from at least 2400 BCE in Africa and Asia (Walters 1989). Wild melon populations still occur in desert and savanna regions of Africa, Arabia, southwestern-Asia and Australia (Sauer 1993; Zohary and Hopf 1993). The cultivation of melon in Europe started only in the 13th CENT (Stepansky *et al.* 1999). The Hungarians already named melon (*dinnye*) as *dinna* in the early 11th CENT (Szamota and Zolnai 1902-1906), which indicates an early use of melon in the Middle Ages of Central Europe, so the samples of the present chapter recovered from the 15th CENT Budapest may originate from one of the oldest cultivated melons in Europe.

MATERIALS AND METHODS

Seed Samples

Seed remains of medieval melon (*C. melo*) were excavated from a 15th CENT sediment layer in a well at the King's Palace of Árpád Dynasty (8th well, Mansion Teleki), Buda Hill, Budapest (Figure 1. *a*), Hungary (F Gyulai et al. this volume; Gyulai *et al.* 2006). In total, 28,117 melon seeds were recovered (Figure 1. *b*) as the well was apparently used as dust hole in the Middle Ages. Seeds were selected (Figure 1. *c*) and surface sterilized by NaOCL (8 %) for 20 min, and incubated on aseptic tissue culture medium (Figure 1. *d*) for a month, according to Gyulai *et al.* (2003).

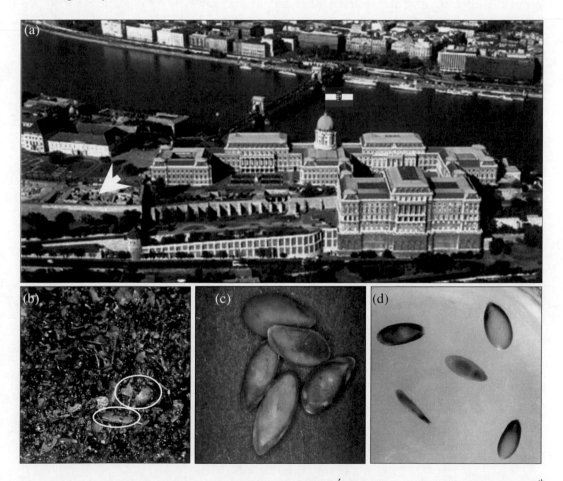

Figure 1. Excavations and seed remains. *(a)* King's Palace of Árpád Dynasty (Budapest, Hungary, 13th CENT ca 1243) with the 15th CENT site (arrowhead). *(b)* Excavated seeds after wet sieving and floatation including melon (*Cucumis melo*) seeds *(c)* before and *(d)* after surface sterilization and incubation on aseptic agar media in a germination trial.

Elimination of Contamination

Ancient seeds were incubated for a month in aseptic tissue culture medium (Figure 1. *d*) to eliminate contamination before DNA extraction, prior to surface sterilization according to Gyulai *et al.* (1992). Exogenously and endogenously contaminated seeds infected by fungi and bacteria were eliminated from further analyses.

Comparative aDNA Analysis with Current Melon Varieties

Seeds of current melon cultivars (1 to 47) were selected from the gene bank of the Agrobotanical Institute Tápiószele (ABI, Hungary) and sown in compost in glass houses. Seedlings at the 2-3 leaf stage were transplanted and grown in 5 m x 5 m field plots in duplicate experiments. Rind, flash and seed morphology was analyzed at ripening time according to Szabó *et al.* (2005).

Medieval seeds were ground in an aseptic mortar individually with liquid nitrogen in a laminar air flow cabinet of archaeogenetic lab of the St Stephanus University, Gödöllő, Hungary. The aDNA samples were extracted by the CTAB (cethyltrimethylammonium bromide) method according to Murray and Thompson (1980), Doyle and Doyle (1990) followed by an RNase-A (from bovine pancreas) treatment (Sigma, R-4875), for 30 min at 37° C. Seed-DNA of current cultivars was also extracted in CTAB buffer, followed by an RNase-A treatment (Sigma, R-4875) for 30 min at 37° C in each case. The quality and quantity of extracted DNA (Figure 2) were measured (2 µl) by a NanoDrop ND-1000 UV-Vis spectrophotometer (NanoDrop Technologies, Delaware, USA – BioScience, Budapest, Hungary). DNA samples were adjusted to concentration of 30 ng/µl with ddH$_2$O and subjected to PCR amplification (Szabó *et al.* 2008a,b).

PCR

A minimum of two independent DNA preparations of each sample were used for PCR amplification following the basic protocols of Gyulai *et al.* (2006). A negative control which contained all the necessary PCR components except template DNA was included in PCR runs. Pooled DNA of medieval melon extracted from the non infected seeds was used according to Michelmore *et al.* (1991).

ITS

For *ITS analysis,* a medieval melon sample was compared to current melon varieties using the primer pair 5'-*tcg taa caa ggt ttc cgt agg tg*-3' and 5'-*tcc tcc gct tat tga tat gc*-3' complementary to the evolutionary conserved spacer regions of the nuclear ribosomal (rDNA) gene cluster including ITS1-5.8S-ITS2 (*i*nternal *t*ranscribed *s*pacer) according to Hsiao *et al.* (1995) and Jobst *et al.* (1998).

SSR

Melon-specific microsatellites were amplified at eight *Cm* (*Cucumis melo*) loci. Names of the microsatellites refer to species name (*Cm*), repeat unit (e.g. TC) and clone number (e.g. 13) as in *Cm*TC13. The eight effective SSR primer pairs (with code, primer sequences and target sequences) were: (1) *Cm*CT44 (*tca act gtc cat ttc tcg ctg and ccg taa aga cga aaa ccc ttc*: $[(ct)_{10}tgtt(ct)_3]$); (2) *Cm*AG59 (*ttg ggt ggc aat gag gaa and ata tga tct tcc att tcc a*: $[(ga)_2a(ag)_8]$); (3) *Cm*GA104 (*tta ctg ggt ttt gcc gat tt and aat tcc gta ttc aac tct cc*: $[(ga)_{14}aa(ga)_3]$); (4) *Cm*TA134a (*acg tgc ttc agt aaa cat g and ccg aca ttg aaa acc aac ttc*: $[(ta)_{12}]$); (5) *Cm*CT134b (*gct cct cct taa ctc tat ac and gca tta tta ccc atg tac gag*: $[(ta)_2(ct)_8(at)_7]$); (6) *Cm*CTT144 (*caa aag gtt tcg att ggt ggg and aaa tgg tgg ggg ttg aat agg*: $[(ctt)_{10}ctac(ctt)_4]$); (7) *Cm*TC168 (*atc att gga tgt ggg att ctc and aca gat gga tga aac ctt agg*: $[(tc)_{14}]$) and (8) *Cm*CT170b (*att gcc caa cta aac taa acc and cac aac aca ata tca tcc ttg*: $[(ct)_8]$) (Katzir *et al.* 1996; Danin-Poleg *et al.* 2001).

ALF

Microsatellite fragments were forwarded for ALF analysis using ALF ExpressII (Pharmacia – Amersham, AP-Hungary, Budapest, Hungary). One strand of each of the SSR primer pairs was labeled with Cy5 dye. PCR and ALF analysis were carried out by the protocol of Röder *et al.* (1998), Huang *et al.* (2002) and Gyulai *et al.* (2006).

RAPD

For RAPD analysis, 60 primers from primer sets OP-A, K and -M, (1 to 20 in each case) were applied (Operon Technology, USA) according to Williams *et al.* (1990).

Sequencing

Amplified fragments were isolated from the agarose gel with a spin column (Sigma, 56501) and subjected to automated fluorescent DNA sequencer (ABI PRISM 3100 Genetic Analyzer, Applied Biosystems, Hungary). Authenticity of aDNA sequences were proved by BLAST (NCBI) analysis in each case.

Sequence Analysis

Sequence alignments were carried out by BioEdit Sequence Alignment Editor (North Carolina State University, USA) (Hall 1999), MULTALINE (Combet *et al.* 2000) and CLUSTAL W (Thompson *et al.* 1994) programs. For SSR, repeat search of FastPCR Program was used (Kalendar *et al.* 2009). BLAST (*B*asic *L*ocal *A*lignment *S*earch *T*ool) analysis was carried out by computer program of NCBI (*N*ational *C*enter for *B*iotechnology *I*nformation, Bethesda, Maryland, USA) (Altschul *et al.* 1997).

Data Analysis

Cluster analysis was carried out by the SPSS-11 program package using the Jaccard Similarity Index (Jaccard 1908) (Average Linkage, within group) based on the presence versus absence of SSR alleles.

The 47 current melon cultivars (designated according to melon types: C - cantalupensis, I – inodorus, R – reticulates) are listed in Figure 5b (Szabó et al. 2005).

CONCLUSIONS

Genus *Cucumis*, based on binomial classifications, include 9 (Linnaeus 1753; Kirkbride 1993) to 13 (Naudin 1859) and 34 (Andres 2004) species. Two of them, melon (*C. melo*) and cucumber (*C. sativus*), have worldwide distribution and economical importance. Species melon (*C. melo*, 2n = 24) (Dane and Tsuchiya 1976), based on trinomial classification, has two subspecies (*ssp.*) as melon (*C. melo* ssp. *melo*) and the wild melon (*C. melo* ssp. *agrestis*) (Jeffrey 1990; Munger and Robinson 1991). Melon (*C. melo* ssp. *melo*) are divided into further interfertile cultivar groups such as *reticulatus, cantalupensis, inodorus, flexuosus, conomon, chito, dudaim* and *makuwa* (Pitrat et al. 2000; Janick and Paris 2006). Since melon (*C. melo* ssp. *melo*) shows extreme variation (Liu et al. 2004), especially in vegetative and fruit morphology, tools of molecular markers have been used for genotype identification such as RAPD (Staub et al. 2004; Zhuang et al. 2004; López-Sesé et al. 2003; Stepansky et al. 1999; Silberstein et al. 1999), AFLP (Wang et al. 1997; Périn et al. 2002), QTL and SNPs (Monforte et al. 2004; Morales et al. 2004), organelle DNA (Perl-Treves and Galun 1985; Helm and Helmleben 1997), and ITS of rDNA (Garcia-Mas et al. 2004; Jobst et al. 1998; Kavanagh and Timmins 1988; Hsiao et al. 1995; Szabó et al. 2005).

Microsatellites (SSR) has also provided species-specific allele patterns in melon (Katzir et al. 1996; Garcia-Mas et al. 2004; Danin-Poleg et al. 2001; Oliver et al. 2001; Ritschel et al. 2004).

aDNA Extracted

The preselected and surface sterilized medieval seeds were incubated on aseptic tissue culture medium F6 (Figure 1. *d*) for a month in a failed germination trial (Gyulai et al. 2001). These seeds provided an optimal aseptic source for aDNA extraction. Possible DNA contamination from the laboratory investigations was also excluded by this aseptic treatment

aDNA from non-infected medieval seeds (Figure 1. *d*) was extracted individually and analyzed on agarose gel (0.8 %) which showed different levels of degradation compared to current melons (Figure 3. *a* and *b*). However, the amount of aDNA was higher in each case than the 1-5 ng / 10 µl detection limit of ethidium bromide staining (Sambrook et al. 1989).

The quantity and quality of aDNA compared to current melon DNAs showed good values in ratios of $A_{260/280}$ and $A_{260/230}$ (Figure 2).

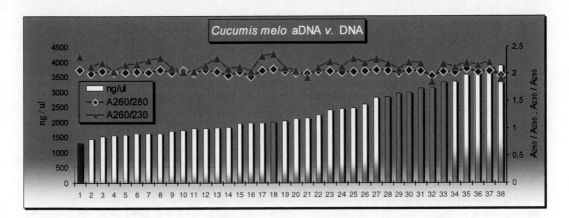

Figure 2. Quantity (0 – 900 ng/µl) **and quality** (0 – 2.5 ratios of UV Absorbances at 230, 260 and 280 nm) of single seed aDNAs of *(1)* 15th CENT melon (Budapest, Hungary) compared to current melon (*Cucumis melo*) cultivars *(2 - 38)*.

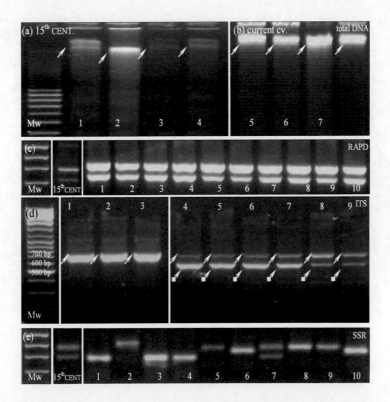

Figure 3. Samples of agarose gel electrophoresis of aDNA. *(a)* Total aDNA (arrows) isolated from single medieval (15th CENT) melon (*Cucumis melo*) seeds (1-4) separated on agarose (0.8 %) gel (Mw - 100 bp ladder from 400 bp), and *(b)* compared to current varieties (5-8). *(c)* Samples of monomorphic RAPD (OP/M-16) amplification. *(d)* ITS analysis at ITS1-5.8S-ITS2 locus of nuclear rDNA in single seed remains (*1 to 9*) of 15th CENT seeds without (1 to 3, 667 bp, arrows) and with a contaminating ITS fragment (*4 to 9*, 579 bp) of fungus *Aspergillus nidulans* (**arrow with ■**). *(e)* polymorphic SSR fragments (*Cm*GA104) amplified in the 15th CENT melon compared to current melon cultivars (*1 to 10*) (Mw - DNA ladder 200 to 500 bp).

For safe aDNA analysis, the most important step was to eliminate both the exogenously and endogenously infected seeds, because bacterial and fungal DNA-remains can contaminate the plant DNA being studied (Gyulai *et al.* 2006).

ITS Analysis

A melon specific ITS fragment (667 bp) was amplified in the 15th CENT aDNA and in the current varieties. Several surface sterilized medieval seeds showed endogenous contamination with an extra ITS fragment with 579 bp length (Figure 3. *d. 4 - 9*), which, after sequencing, was found to be an *Aspergillus ustus* (NCBI: AY373879). DNA samples of these contaminated seeds were excluded from further analysis. None of the current varieties showed any ITS contamination. SNP analysis revealed nucleotide changes in the highly conserved ITS1-5.8S-ITS2 non-coding region of rDNA at the 94-95 bp (GC to either RC, AC, AG, AS, GG, GS, or RS) of ITS1; and at 414th bp (A-to-T substitution), 470th bp (T to Y or C), 610th bp (A to R or G) and 633rd bp (A-to-G transition) of ITS2 (Figure 4).

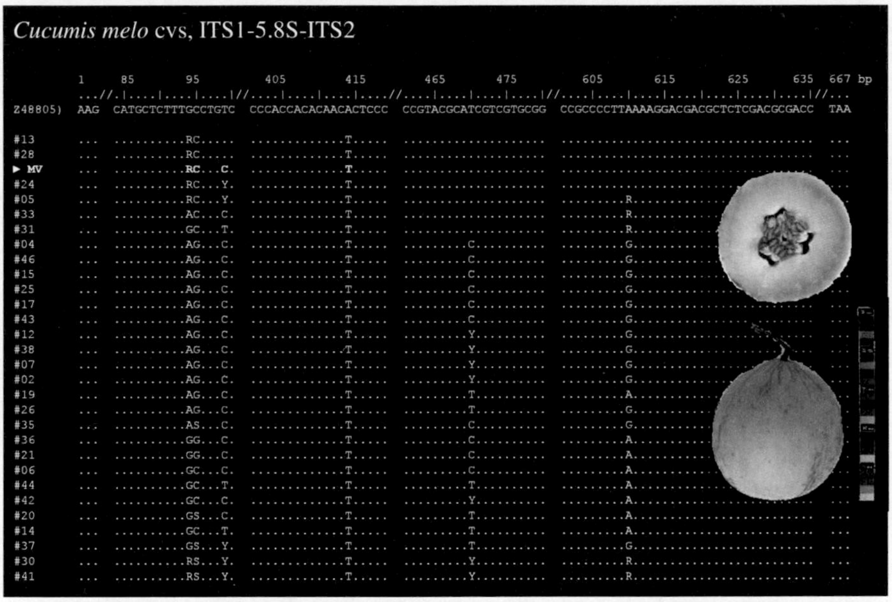

Figure 4. Consensus sequence alignments of the highly conserved ITS1-5.8S-ITS2 fragments (Figure 3. *d*) of nuclear rDNA (*1* to *667 bp*) of 15th CENT melon (▶ MV) compared to NCBI data base (#Z48805, *Cucumis melo*) and current melon (*Cucumis melo*) cultivars and landraces (R - A or G, S - G or C; Y - T or C; dots indicate consensus sequences). Images of current cv. *Hógolyó* (# 24, *Snowball*) (*color bar 25 cm*), an *inodorus* type winter melon, provided morphological reconstruction to the 15th CENT melon showing the closest sequence similarity.

ALF-SSR Analysis

Eight of the twenty SSR primers amplified a total of 463 alleles at 40 loci among the 47 melon cultivars and the medieval sample.

Amplified SSR loci included *Cm*CT44 (2 alleles), *Cm*AG59 (5 alleles), *Cm*GA104 (5 alleles) (Figure 3. *e*), *Cm*CT134 (4 alleles), *Cm*TA134 (6 alleles), *Cm*TC168 (6 alleles), *Cm*CT170 (5 alleles) (Figure 5a) and *Cm*CTT144 (7 alleles) (Figure 5b).

The number of alleles per marker ranged from 2 to 7 with an average of 5.7, which was similar to the range (3 to 7), observed by Garcia-Mas *et al.* (2004) for melon specific markers. However, orthologous microsatellites can appear at higher allele numbers (9 to 15) in species of C. *melo* and C. *sativus* (Garcia-Mas *et al.* 2004).

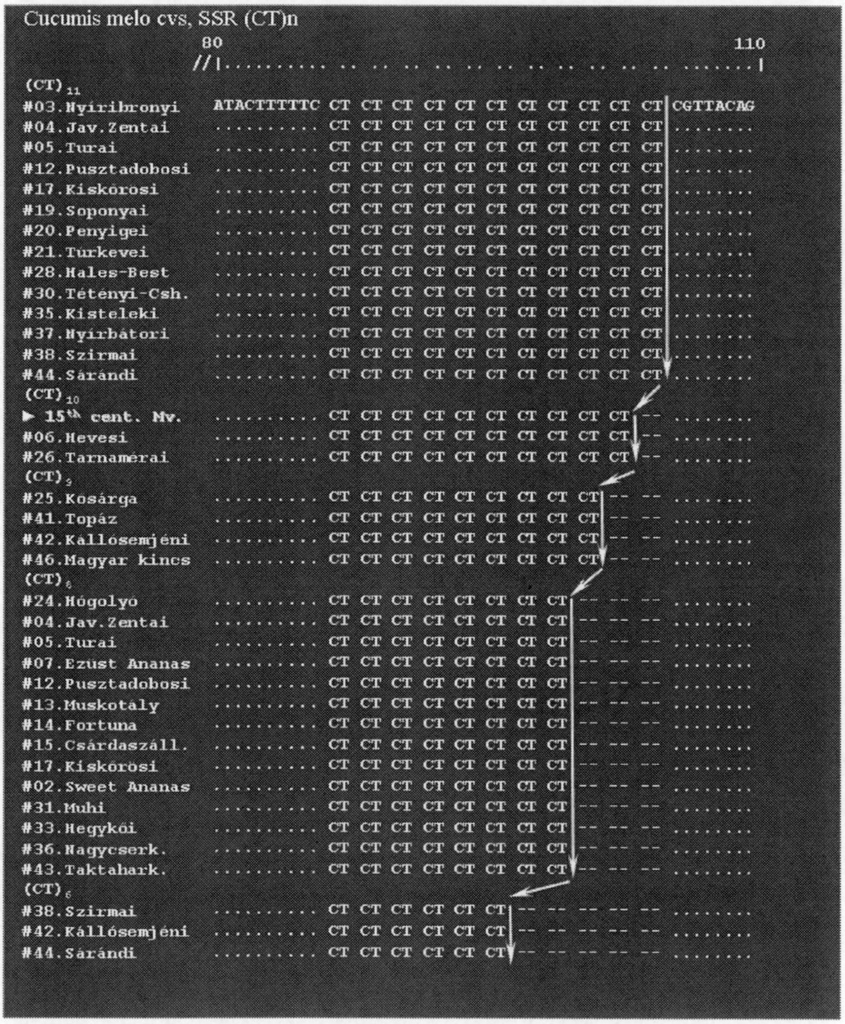

Figure 5a. Consensus sequence alignments of dinucleotide microsatellite DNA fragments at the (CT)n locus (80-110 bp of *Cm*CT170, Katzir *et al.* 1996, Danin-Poleg *et al.* 2001, Szabó *et al.* 2005) in the medieval (▶MV) from the 15th CENT and current melon cultivars (*Cucumis melo*). Arrows indicate the shortening length of SSRs.

Sequence analysis of microsatellites at the loci of *Cm*CT170 (5 alleles) (Figure 5a) and *Cm*CTT144 (7 alleles) (Figure 5b) revealed that the length of SSRs did not show a time-dependent variation in lengths as a result of microevolution, as medieval melon showed SSRs of intermediate sizes compared to current cultivars at both SSR loci.

```
Cucumis melo cvs, SSR (CT)n
                    80                                                    110
                    //|..........  ..  ..  ..  ..  ..  ..  ..........|
(CT)11
#03.Nyiribronyi  ATACTTTTTC CT CT CT CT CT CT CT CT CT CT CGTTACAG
#04.Jav.Zentai   .......... CT CT CT CT CT CT CT CT CT CT CT ........
#05.Turai        .......... CT CT CT CT CT CT CT CT CT CT CT ........
#12.Pusztadobosi .......... CT CT CT CT CT CT CT CT CT CT CT ........
#17.Kiskőrösi    .......... CT CT CT CT CT CT CT CT CT CT CT ........
#19.Soponyai     .......... CT CT CT CT CT CT CT CT CT CT CT ........
#20.Penyigei     .......... CT CT CT CT CT CT CT CT CT CT CT ........
#21.Túrkevei     .......... CT CT CT CT CT CT CT CT CT CT CT ........
#28.Hales-Best   .......... CT CT CT CT CT CT CT CT CT CT CT ........
#30.Tétényi-Csh. .......... CT CT CT CT CT CT CT CT CT CT CT ........
#35.Kisteleki    .......... CT CT CT CT CT CT CT CT CT CT CT ........
#37.Nyírbátori   .......... CT CT CT CT CT CT CT CT CT CT CT ........
#38.Szirmai      .......... CT CT CT CT CT CT CT CT CT CT CT ........
#44.Sárándi      .......... CT CT CT CT CT CT CT CT CT CT CT ........
(CT)10
▶ 15th cent. Mv. .......... CT CT CT CT CT CT CT CT CT CT --......
#06.Hevesi       .......... CT CT CT CT CT CT CT CT CT CT --......
#26.Tarnamérai   .......... CT CT CT CT CT CT CT CT CT CT --......
(CT)9
#25.Kósárga      .......... CT CT CT CT CT CT CT CT CT -- -- ........
#41.Topáz        .......... CT CT CT CT CT CT CT CT CT -- -- ........
#42.Kállósemjéni .......... CT CT CT CT CT CT CT CT CT -- -- ........
#46.Magyar kincs .......... CT CT CT CT CT CT CT CT CT -- -- ........
(CT)8
#24.Hógolyó      .......... CT CT CT CT CT CT CT CT -- -- -- ........
#04.Jav.Zentai   .......... CT CT CT CT CT CT CT CT -- -- -- ........
#05.Turai        .......... CT CT CT CT CT CT CT CT -- -- -- ........
#07.Ezüst Ananas .......... CT CT CT CT CT CT CT CT -- -- -- ........
#12.Pusztadobosi .......... CT CT CT CT CT CT CT CT -- -- -- ........
#13.Muskotály    .......... CT CT CT CT CT CT CT CT -- -- -- ........
#14.Fortuna      .......... CT CT CT CT CT CT CT CT -- -- -- ........
#15.Csárdaszáll. .......... CT CT CT CT CT CT CT CT -- -- -- ........
#17.Kiskőrösi    .......... CT CT CT CT CT CT CT CT -- -- -- ........
#02.Sweet Ananas .......... CT CT CT CT CT CT CT CT -- -- -- ........
#31.Muhi         .......... CT CT CT CT CT CT CT CT -- -- -- ........
#33.Hegykői      .......... CT CT CT CT CT CT CT CT -- -- -- ........
#36.Nagycserk.   .......... CT CT CT CT CT CT CT CT -- -- -- ........
#43.Taktahark.   .......... CT CT CT CT CT CT CT CT -- -- -- ........
(CT)6
#38.Szirmai      .......... CT CT CT CT CT CT -- -- -- -- -- ........
#42.Kállósemjéni .......... CT CT CT CT CT CT -- -- -- -- -- ........
#44.Sárándi      .......... CT CT CT CT CT CT -- -- -- -- -- ........
```

Figure 5b. Consensus sequence alignments of trinucleotide microsatellite DNA fragments at the (CTT)n locus (45-127 bp of *Cm*CTT144; Katzir *et al.* 1996, Danin-Poleg *et al.* 2001, Szabó *et al.* 2005, Mansour *et al.* 2009) in the medieval (▶MV) from the 15th CENT and current melon cultivars (*Cucumis melo*). Arrows indicate the shortening length of SSRs (cultivar names are listed in Figure 5a).

In certain cases, SSR loci show homoplasy, as the number of core sequence unit can increase or decrease reversibly, making it impossible to distinguish ancestral and derived states (Orti *et al.* 1997; Schulman 2007).

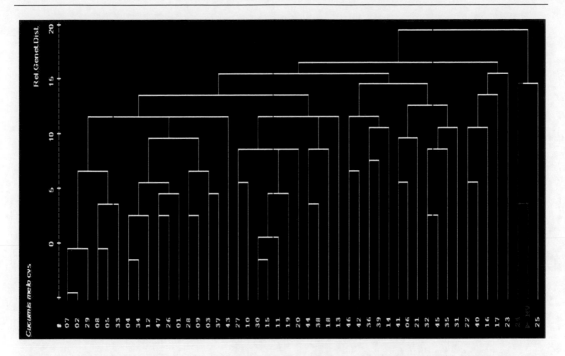

Figure 6. Molecular (SSR) dendrogram of medieval *(►MV)* and modern melon (*Cucumis melo*) cultivars and landraces (*1* to *47*) (listed in Figure 5a) based on microsatellite DNA fragments diversity. The medieval melon from 15th CENT showed the closest genetic similarity to a registered melon cv. *Hógolyó* (# *24*).

SSR loci can also show autonomous ('*selfish*') life with periods of 'birth' (Messier *et al.* 1996) and 'death' (Taylor *et al.* 1999). However, SSR loci provided stable molecular markers in wheat (Röder *et al.* 1998), and genetic tools for tracing directional evolution in maize (Vigouroux *et al.* 2003).

Similar to ITS sequence analysis (Figure 4) cluster analysis (Figure 6) based on the presence versus absence of microsatellite alleles also showed that the 15th CENT melon had the closest SSR similarity to the current melon cv. '*Hógolyó*' ('*snowball*') selected from an old Hungarian melon landrace, even the case that in the two sequenced microsatellites the 15th CENT melon belonged in different subgroup (Figure 5a and 5b). Melon cv. #24 *Hógolyó* ('snowball'), registered in 1981 (OMMI 2004), was selected from old populations of Hungarian landrace, which has apparently showed ancient characteristics of indeterminate growing habit and vine development in field test.

RAPD-PCR

To test aDNA for PCR amplification capacity, RAPD analysis was also applied with 60 primers from primer sets OP-A, -K and -M, (1 to 20 in each case) (Operon Technology, USA) resulting in a total of 105 fragments in the medieval and current melons. Of them, 61 were monomorphic and 44 showed polymorphisms (Figure 3. *c*).

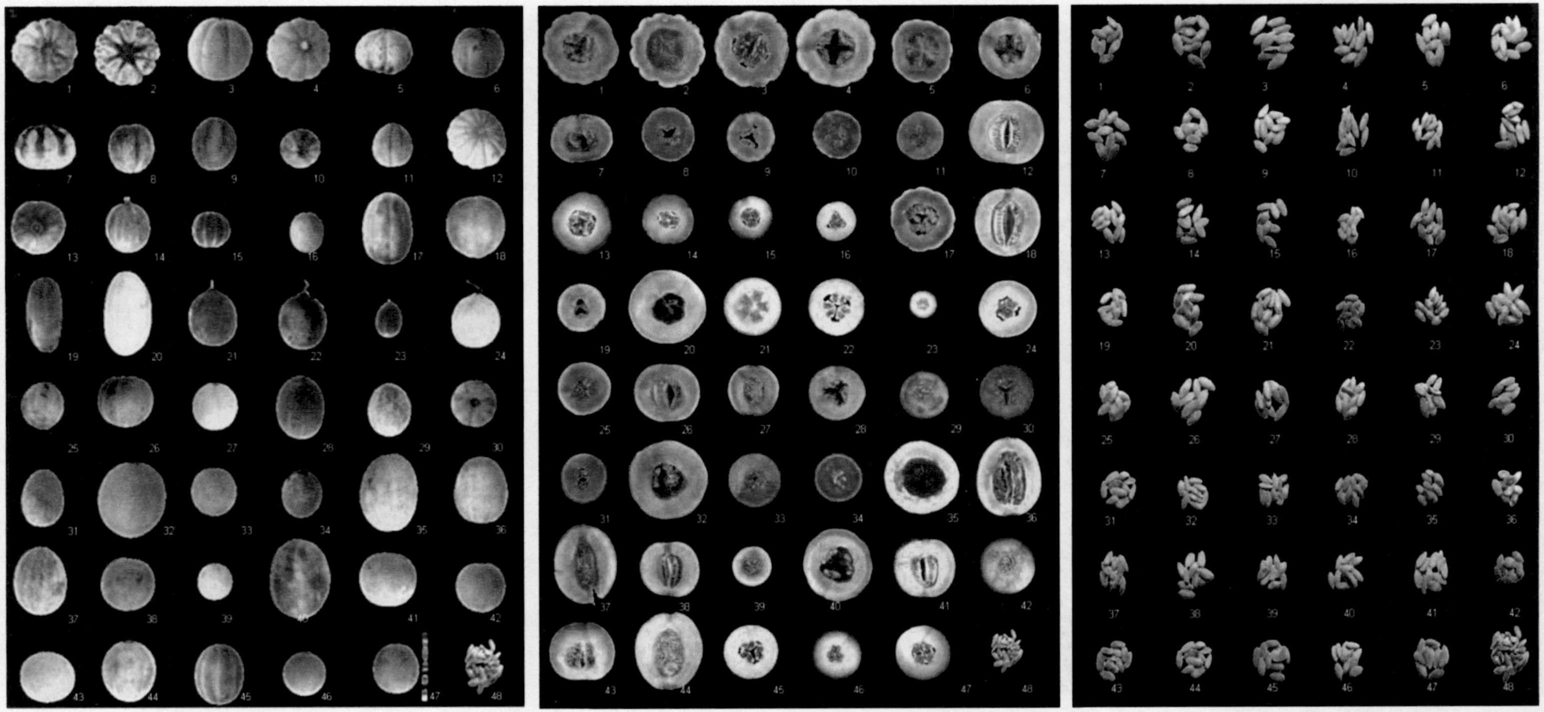

Figure 7. Morphological reconstruction of medieval melon. Rind (*a*), flash (*b*) and seed (*c*) types of current melons (*Cucumis melo*) at ripening time used for morphological reconstruction of 15th CENT (Budapest, Hungary) melon (*48*). Cultivar names are listed in Figure 5b. Color tool bar (*a*) indicates 25 cm. Molecular reconstruction revealed that medieval melon had the closest genetic similarity to a registered melon cv. *Hógolyó* (# 24, '*Snowball*') of *inodorus* type winter melon with smooth yellow rind and green flesh color.

The level of RAPD polymorphism was similar to an analysis in which 145 of the 500 RAPD fragments were found to be polymorphic, among 93 F_2 individuals originating from current melon cultivars (Oliver *et al.* 2001).To conclude, DNA sequence analyses of the evolutionary conserved ITS1-5.8S-ITS2 regions of nuclear rDNA and the nuclear SSR fragments diversity indicate the importance of tools of molecular archaeogenetics in genotype reconstruction and resurrection of ancient and extinct plant lines and varieties.

REFERENCES

Altschul SF, TL Madden, AA Schaffer, JH Zhang, Z Zhang, W Miller, DJ Lipmand (1997) Gapped BLAST and PSI-BLAST: a new generation of protein database search programs. *Nucleic Acids Research* 25: 3389-3402.

Andres TC (2004) Web site for the plant family *Cucurbitaceae* & home of The Cucurbit Network. *http://www.cucurbit.org/family.html.*

Combet C, C Blanchet, C Geourjon, G Deléage (2000) NPS@: Network Protein Sequence Analysis. *Trends in Biochemical Sciences* 25(3): 147-150.

Dane F, T Tsuchiya (1976) Chromosome studies in the genus *Cucumis*. *Euphytica* 25: 367-374.

Danin-Poleg Y, N Reis, G Tzuri, N Katzir (2001) Development and characterization of microsatellite markers in *Cucumis*. *Theor Appl Genet* 1002: 61-72.

Doyle JJ, JL Doyle (1990) Isolation of plant DNA from fresh tissue. *Focus* 12: 13-15.

Garcia-Mas J, AJ Monforte, P Arús (2004) Phylogenetic relationships among *Cucumis* species based on the ribosomal internal transcribed spacer sequence and microsatellite markers. *Plant Syst Evol* 248: 191-203.

Gorman CF (1969) Hoabinhian: a people-toll complex with early plant associations in southeast Asia. *Science* 163: 671-673.

Gyulai G, J Janovszky, E Kiss, L Lelik, A Csillag, LE Heszky (1992) Callus initiation and plant regeneration from inflorescence primordia of the intergeneric hybrid *Agropyron repens* (L.) Beauv. x *Bromus inermis* Leyss. cv. *nanus* on a modified nutritive medium. *Plant Cell Rep* 11: 266-269.

Gyulai G, A Magda, J Kiss, F Gyulai, L Holly, L Heszky (2001) DNA isolation and PCR amplification from 700-year-old seed remains (DNS izolálás és PCR-amplifikáció 700 éves növénymagvakból). 7th Annual Meeting of Hungarian Plant Breeders, p.89.

Gyulai G, Z Mester, J Kiss, L Szemán, L Heszky, A Idnurm (2003) Somaclone breeding of reed canarygrass (*Phalaris arundinacea* L). *Grass Forage Sci* 58: 210-215.

Gyulai G, M Humphreys, R Lagler, Z Szabó, Z Tóth, A Bittsánszky, F Gyulai, L Heszky (2006) Seed remains of common millet from the 4th (Mongolia) and 15th (Hungary) centuries: AFLP, SSR and mtDNA sequence recoveries. *Seed Science Research* 16: 179-191.

Hall TA (1999) BioEdit: a user-friendly biological sequence alignment editor and analysis program for Windows 95/98/NT. *Nucleic Acids Symposium* Series No. 41: 95-98.

Helm MA, V Helmleben (1997) Characterization of new prominent satellite DNA of *Cucumis metuliferus* and differential distribution of satellite DNA in cultivated and wild species of *Cucumis* and related genera of *Cucurbitaceae*. *Euphytica* 94: 219-226.

Hsiao C, NJ Chatterton, KH Asay, KB Jensen (1995) Phylogenetic relationships of the monogenomic species of the wheat tribe, *Triticeae* (*Poaceae*), inferred from nuclear rDNA (internal transcribed spacer) sequences. *Genome* 38: 211-223.

Huang XQ, A Börner, MS Röder, MW Ganal (2002) Assessing genetic diversity of wheat (*Triticum aestivum* L.) germplasm using microsatellite markers. *Theor Appl Genet* 105: 699-707.

Jaccard P (1908) Nouvelles recherches sur la distribution florale. *Bull Soc Vaud Sci Nat* 44: 223-270.

Janick J, HS Paris (2006) The Cucurbit Images (1515-1518) of the Villa Farnesina, Rome. *Annals of Botany* 97: 165-176.

Jeffrey D (1990) Appendix: An outline classification of the *Cucurbitaceae*. In: Bates DM, Robinson RW, Jeffrey C Biology and utilization of the *Cucurbitaceae*. pp.449-463. 485p. Ithaca and London, Cornell University.

Jobst J, K King, V Hemleben (1998) Molecular evolution of the Internal Transcribed Spacers (ITS1 and ITS2) and phylogenetic relationships among species of the family Cucurbitaceae. *Mol Phylogenet Evol* 9: 204-219.

Katzir N, Y Danin-Poleg, G Tzuri, Z Karchi, U Lavi, PB Cregan (1996) Length polymorphism and homologies of microsatellites in several *Cucurbitaceae* species. *Theor Appl Genet* 93: 1282-1290.

Kalendar R, D Lee, AH Schulman (2009) FastPCR Software for PCR Primer and Probe Design and Repeat Search. *Genes, Genomes and Genomics*, 3(1): 1-14. [http://primerdigital.com/fastpcr.html].

Kavanagh TA, JN Timmins (1988) Structure of melon rDNA and nucleotide sequence of the 17-25S spacer region. *Theor Appl Genet* 76: 673-680.

Keimer L (1924) Die Gartenpflanzen im alten Ägypten. (1967 reprint). Vol. 1, Hamburg/Berlin, 117 p.

Kirkbride JH (1993) Biosystematic monograph of the genus *Cucumis* (*Cucurbitaceae*). Parkway Publishers. NC, USA.

Kroll H (1982) Kulturpflanzen von Tiryns. Archaeologischer Anzeiger, 1982: 467-485.

Kroll H (1999) Literature on archaeological remains of cultivated plants (1997/1998). Veget Hist Archaeobot 8: 129-163.

Kroll H (2000) Literature on archaeological remains of cultivated plants (1998/1999). *Veget Hist Archaeobot* 9: 31-68.

Linnaeus C (1753) Species Plantarum. 2 Vols. Salvius, Stockholm. Facsimile edition (1957-1959), Ray Society, London.

Liu L, F Kakihara, M Kato (2004) Characterization of six varieties of *Cucumis melo* L. based on morphological and physiological characters, including shelf-life of fruit. *Euphytica* 135: 305-313.

López-Sesé AI, JE Staub, ML Gómez-Guillamón (2003) Genetic analysis of Spanish melon (*Cucumis melo* L.) germplasm using a standardized molecular-marker array and geographically diverse reference accessions. *Theor Appl Genet* 108: 41-52.

Mansour A, JAT da Silva, G Gyulai (2009) Assessment of Molecular (Dis)similarity: The Role of Multiple Sequence Alignment (MSA) Programs in Biological Research. *Genes, Genomes and Genomics* 3:23-30.

Messier W, S Li, C Stewart (1996) The birth of microsatellites. *Nature* 381:483.

Michelmore RW, I Paran, RV Kesseli (1991) Identification of markers linked to disease-resistance genes by bulked-segregant analysis: a rapid method to detect markers in specific genomic regions by using segregating populations. *Proc Natl Acad Sci USA* 88: 9828-9832.

Morales M, E Roig, AJ Monforte, P Arús, J Garcia-Mas (2004) Single-nucleotide polymorphisms detected in expressed sequence tags of melon (*Cucumis melo* L.). *Genome* 47: 352-360.

Monforte AJ, M Oliver, MJ Gonzalo, JM Alvarez, R Dolcet-Sanjuan, P Arús (2004) Identification of quantitative trait loci involved in fruit quality traits in melon (*Cucumis melo* L.). *Theor Appl Genet* 108: 750-758.

Munger HM, RW Robinson (1991) Nomenclature of *Cucumis melo* L. *Cucurbit Genet Coop Reports* 14: 43-44.

Murray MG, WF Thompson (1980) Rapid isolation of high molecular weight plant DNA. *Nucleic Acids Res* 8: 4321-4325.

Naudin CV (1859) Essais d'une monofraphie des especè et des varieties du genre *Cucumis*. *Ann. Sci. Nat. Bot.* Sér 4, 11:5-87.

Oliver JL, J Garcia-Mas, M Cardús, N Pueyo, AI López-Sesé, M Arroyo, H Gómez-Paniagua, P Arús, CM de Vicente (2001) Construction of a reference linkage map of melon. *Genome* 44: 836-845.

OMMI (2004) National list of varieties. ed. K Neszmélyi. National Institute for Agricultural Quality Control, Budapest, Hungary.

Orti G, DE Pearse, JC Avise (1997) Phylogenetic assessment of length variation at a microsatellite locus. *Proc Natl Acad Sci USA* 94: 10745-10749.

Perl-Treves R, E Galun (1985) The *Cucumis* plastome: physical map, intrageneric variation and phylogenetic relationships. *Theor Appl Genet* 71: 417-429.

Périn C, LS Hagen, V De Conto, N Katzir, Y Danin-Poleg, V Portnoy, S Baudracco-Arnas, J Chadoeuf, C Dogimont, M Pirat (2002) A reference map of *Cucumis melo* based on two recombinant inbred line populations. *Theor Appl Genet* 104: 1017-1034.

Pitrat M, P Hanelt, K Hamer (2000) Some comments on infraspecific classification of cultivars of melon. *Acta Hort* 510: 29-36.

Ritschel PS, TC Lins, RL Tristan, GSC Buso, JA Buso, ME Ferreira (2004) Development of microsatellite markers from an enriched genomic library for genetic analysis of melon (*Cucumis melo* L). *BMC Plant Biol* 4:9.

Röder MS, V Korzun, K Wendehake, J Plaschke, MH Tixier, P Leroy, MW Ganal (1998) A microsatellite map of wheat. *Genetics* 149: 2007-2023.

Sambrook J, EF Fritsch, T Maniatis (1989) Molecular Cloning. A laboratory manual 1-3. 2nd Edition, Cold Spring Harbor Lab Press, New York.

Sauer JD (1993) Historical geography of crop plants - a select roster. CRC Press, Boca Raton, Florida. pp.1-308.

Schulman AH (2007) Molecular markers to assess genetic diversity. *Euphytica* 158: 313-321.

Silberstein L, I Kovalski, R Huang, K Anagnostou, MMK Jahn, R Perl-Treves (1999) Molecular variation in melon (*Cucumis melo* L.) as revealed by RFLP and RAPD markers. *Sci Hort* 79: 101-111.

Szabó Z, G Gyulai, M Humphreys, A Bittsánszky, F Gyulai, R Lagler, J Kiss[1], L Horvath, L Heszky (2004) aDNA analysis of cantaloupe (*Cucumis melo* L) from the Middle Ages compared to modern varieties. In: J Vollmann, H Grausgruber, P Ruckenbauer (eds)

Genetic variation for plant breeding, pp.97-101. The 17th Eucarpia General Meeting, 8-11, September, Tulln, Austria, ISBN 3-900962-56-1.

Szabó Z, G Gyulai, M Humphreys, L Horváth, A Bittsánszky, R Lágler, L Heszky (2005) Genetic variation of melon (*C. melo*) compared to an extinct landrace from the Middle Ages (Hungary) I. rDNA, SSR and SNP analysis of 47 cultivars. *Euphytica* 146: 87-94.

Szabó Z, G Gyulai, Z Tóth, L Heszky (2008a) Morphological and molecular diversity of 47 melon (*Cucumis melo*) cultivars compared to an extinct landrace excavated from the 15th Century. In: *Cucurbitaceae* 2008, Proceedings of the IXth EUCARPIA meeting on genetics and breeding of Cucurbitaceae, Pitrat M (ed) INRA, Avignon (France), May 21-24th, pp.313-321.

Szabó Z, G Gyulai, Z Tóth, A Bittsánszky, L Heszky (2008b) Sequence diversity at the loci of nuclear SSRs and ITS1-5.8S-ITS2 of rDNA of 47 melon (*Cucumis melo*) cultivars and an extinct landrace excavated from the 15th century. General Meeting EUCARPIA, Valencia, Spain, pp. 244-249.

Staub JE, AI López-Sesé, N Fanourakis (2004) Diversity among melon landraces (*Cucumis melo* L.) from Greece and their genetic relationships with other melon germplasm of diverse origins. *Euphytica* 136: 151-166.

Stepansky A, I Kovalski, R Perl-Treves (1999) Intraspecific classification of melons (*Cucumis melo* L.) in view of their phenotypic and molecular variation. *Plant Syst Evol* 217: 313-333.

Szamota I, Gy Zolnai (1902-1906) (reprint 1984) *Magyar oklevél-szótár*. Régi oklevelekben és egyéb iratokban előforduló magyar szók gyűjteménye. Pótlék a Magyar nyelvtörténeti szótárhoz. Budapest. 1902-1906, Hornyánszky. XXXI p., 1210 (OklSz).

Taylor JS, JMH Durkin, F Breden (1999) The Death of a Microsatellite: A Phylogenetic Perspective on Microsatellite Interruptions. *Mol Biol Evol* 16: 567-572.

Thompson JD, DG Higgins, TJ Gibson (1994) CLUSTAL W: improving the sensitivity of progressive multiple sequence alignment through sequence weighting, positions-specific gap penalties and weight matrix choice. *Nucleic Acids Research* 22: 4673-4680.

Vigouroux Y, Y Matsuoka, J Doebley (2003) Directional Evolution for Microsatellite Size in Maize. *Mol Biol Evol* 20:1480-1483.

Walters TW (1989) Historical overview on domesticated plants in China with special emphasis on the Cucurbitaceae. *Economic Botany* 43: 297-313.

Wang YH, CE Thomas, RA Dean (1997) A genetic map of melon (*Cucumis melo* L.) based on amplified fragment length polymorphism (AFLP) markers. *Theor Appl Genet* 95: 791-798.

Williams JGK, Kubelik AR, Rafalski KJ, Tingey SV (1990) DNA polymorphisms amplified by arbitrary primers are useful as genetic markers. *Nucleic Acids Res* 18: 6531-6535.

Zhuang FY, JF Chen, JE Staub, CT Qian (2004) Assessment of genetic relationships among *Cucumis* spp. by SSR and RAPD marker analysis. *Plant Breeding* 123: 167-172.

Zohary D, M Hopf (1993) Domestication of plants in the Old World - The origin and spread of cultivated plants in West Asia, Europe, and the Nile Valley. Clarendon Press, Oxford.

In: Plant Archaeogenetics
Editor: Gábor Gyulai, pp. 105-114

Chapter 9

MOLECULAR DOMESTICATION AND PHYLOGEOGRAPHY OF CUCURBIT SPECIES

*Fenny Dane**

Department of Horticulture, Auburn University, AL 36849, USA

ABSTRACT

The Cucurbitaceae family is comprised of several economically important cultivated species including watermelon, melon, cucumber and squash. Members of the family have several unique traits such as lianous structure of the plant body, fleshy fruits and unusual mode of sex determination. Large scale sequencing analysis of many cucurbit species and almost all cucurbit genera indicated that the family probably originated in Asia and spread via transoceanic long distance dispersal to the different continents. While *Cucurbita* species are among the earliest plants domesticated in the New World, the bottle gourd might have been domesticated in Asia first and later in Africa. Watermelon was probably domesticated in Africa, the cucumber in India, while the origin of domestication of melon is still uncertain. Major efforts are underway to sequence several of the economically important cucurbit genomes and functional genomic studies. Together with archaeogenetical data, this will lead to a deeper understanding of the origin, and genetic basis of cucurbit crop domestication.

INTRODUCTION

Principal food crops of the cucurbit or gourd family, the Cucurbitaceae, include several economically important cultivated plants such as watermelon [*Citrullus lanatus* (Thunb.) Matsum. & Nakai], cucumber (*Cucumis sativus* L.), melon (*C. melo* L.), and squash (*Cucurbita pepo* L. and *C. maxima* Duch.). Agricultural production of cucurbits uses 9 million hectares of land and yields 184 million tons of vegetables, fruits and seeds annually

---

* E-mail: danefen@auburn.edu

(http://faostat.fao.org). The Cucurbitaceae family consists of two well-defined subfamilies and eight tribes representing about 130 genera and 800 species (Jeffrey 2001, 2005; Kocyan et al. 2007). Large scale multigene chloroplast DNA sequence analysis of 114 Cucurbitaceae genera and 25% of the 960 species indicate that the family most likely originated in Asia in the Late Cretaceous ca 70 M yrs ago, followed by a repeated spread of lineages into the African, American, and Australian continents via transoceanic long-distance dispersal (Schaefer et al. 2009). Of the 960 species of the Cucurbitaceae family, approximately 40% are endemic in the American continent, while 28% occur in Africa, 26% in Asia, 2% in Australia and 1% in Europe (Schaefer et al. 2009).

Cucurbits are mostly prostrate or climbing herbaceous annuals with coiled tendrils, characterized by unisexual flowers and inferior ovaries, and development of fleshy fruits and unique mode of sex determination. Their genome size has not changed as significantly as other botanical families and is relatively small, with cucumber at $2n = 2x = 14$, 243.5 Mb (Huang et al. 2009); melon, $2n = 2x = 24$, 450 Mb; watermelon $2n = 2x = 22$, 430 Mb and squash and pumpkin, $2n = 2x = 40$, 539Mb (Arumuganathan and Earle 1991). The genome of cucumber was recently sequenced (Huang et al. 2009) resulting in estimates of a total of 26,682 genes, a mean coding sequence size of 1,046 bp and average of 4.39 exons per gene, and absence of whole genome duplication events. The mitochondrial genome (mtDNA) of Cucurbitaceae, however, is very variable with an estimated range of 390 kb in watermelon to 982 kb in Cucurbita pepo and 2,900 kb in melon. The larger Cucurbita mitochondrial genome in comparison to the Citrullus genome reflects the accumulation of large amounts of chloroplast (cpDNA) sequences and short repeat sequences (Alverson et al. 2010). With completion of a BAC-based physical map of the melon genome (Gonzalez et al. 2010) and sequencing of the complete melon genome (melonomics project), melon has the potential of serving as a model for elucidation of key traits in fruit development. Since crop domestication is a process of evolutionary change in genetics of plants brought about by humans, questions related to the origin of plant domesticates and the identity of wild ancestors can then be answered.

ARCHAEOGENETICAL AND PHYLOGENETIC ANALYSIS

Genus *Cucurbita*

The New World genus *Cucurbita* (squashes, pumpkins and gourds) is composed of 12-14 species with overlapping geographical distributions that today extend from northern South America to up into the mid-latitude eastern US (Smith 2006). At least five different species were domesticated and some species were among the earliest plants domesticated in the New World (*C. maxima, C. moschata, C. ficifolia, C. argyrosperma, C. equadorensis, C. pepo*) The genetic documentation of domestication of cucurbits has focused on the changes in the genetic structure of cucurbits such as morphological changes in seed size and reduction in seed coat thickness (Smith 2006) and comparative analysis of modern and ancient domesticates and their wild relatives using neutrally evolving, noncoding loci and organellar genomes. Smith (2006) used increases in size of *C. pepo* seeds recovered from archeological sites as evidence of separate domestication events in the Americas. Current genetic, biogeographical and archaeological studies suggest independent domestication events for the different species (Sanjur *et al.* 2002). The correct identification of wild progenitor species contributes to the development of archaeological markers which can distinguish archaeological remains of wild and domesticates.

Cucurbita pepo includes both free-living populations and cultivated squashes and gourds with a great diversity of forms including acorn, zucchini, scallop, crookneck, fordhook and pumpkin cultivars. Variation in seed morphology traits, seed size and shape, was detected among some wild and cultivated forms. The variation in seed morphology among cultivars did not result in distinct groupings. However, molecular data consistently supported the separation of domesticated *C. pepo* into 2 subspecies: spp *pepo* and spp *ovifera* (L.) Decker (Katzir *et al.* 2000, Sanjur *et al.* 2002; Decker-Walters *et al.* 2002). *pepo* spp *pepo* includes pumpkins, zucchini and other marrow squashes, Mexican landraces and a few ornamental gourds, are known only from cultivated forms and may have been domesticated in southern Mexico (Decker-Walters *et al.* 2002).

C. pepo ssp. *ovifera* compriseof both domesticated, cultivated squash cultivars and most ornamental gourds and can be divided into 2 distinctly molecularly divergent groupings. Wild populations in Texas are classified as var. *texana* (Scheele) Decker, whereas populations outside Texas belong to var. *ozarkana* Decker-Walters (Decker-Walters *et al.* 2002). Thus, molecular genetic data has shown that *C. pepo* was domesticated independently from 2 divergent populations and that *C. pepo* spp *ovifera* var. *ozarkana* is most likely descended from one of those progenitor populations (Emshwiller 2006).

Evidence based on examination and reexamination of early domesticate assemblages from caves in Mexico indicated that *C. pepo* is the earliest domesticate (~10,000 before present, BP) in the Guila Naquitz cave in Oaxaca, Mexico (Smith 1997, 2005). *C. pepo* is also the earliest domesticate in the Coxcalan Cave dated to 7920 years BP. The bottle gourd (*Lagenaria siceraria*) is dated at 7200 years BP, while *Cucurbita argyrosperma* (classified on basis of distinctive morphology) does not appear until 2,065 years BP (Smith, 2005). The four archeological sites that have yielded evidence of indigenous domesticated plants in eastern North America are Phillips Spring, Hayes, Napoleon Hollow and Riverton. There is evidence for cultivation of domesticated bottle gourd (*Lagenaria siceracia*) and the possible cultivation of *C. pepo* squash (Smith and Yarnell, 2009) in eastern North America at 3800 BP.

The earliest archaeological remains indicative of domestication of *Cucurbita moschata* were discovered in southern Mexico (7000 BP) and in coastal Peru (5000 BP) (Decker-Walters and Walters 2000). There is no evidence of the existence of *C. moschata* in Europe before the 17th century. Molecular data, using amplified polymorphic DNA (AFLP) and sequence-related amplified polymorphism (SRAP) markers, supports the existence of two independent domestication events in Central and South America (Ferriol *et al.* 2004).

Mitochondrial DNA was used to study phylogenetic relationships among domesticated and wild species of *Cucurbita*. *C. angyrosperma* was likely domesticated from a wild Mexican gourd, *C. sororia*, in southwest Mexico, and *C. andreana* is thought to the wild progenitor of *C. maxima* (Sanjur *et al.* 2002).

Genus *Citrullus*

The African origin of *Citrullus lanatus,* watermelon, is almost certain because of the diversity of spontaneous forms, the presence of wild related species on the African continent (Pitrat *et al.* 1999; Dane and Liu 2007), and recent biogeographical studies of the African cucurbit genera (Schaefer *et al.* 2009). African Cucurbitoidea (25 genera) are thought to be the result of long-distance dispersals from Asia to Africa and from America to Africa. The watermelon and its sister species *C. colocynthis* are thought to have evolved from a common ancestor as recently as 2 M yrs ago (Schaefer *et al.* 2009).

The earliest reliable finds of seeds from watermelon in the world come from an archaeological site- Uan Muhuggiag in southwest Lybia (5000 BP, Wasylikowa and van der Veen, 2004), in Egypt in the tomb of Tutankhamun (ca 1330 BCE, Hepper 1990) and a temple in Sudan (ca 1500 BCE, van Zeist, 1983), suggesting that a wild form of watermelon was available in Libyan Sahara 5000 years ago and domestication of the species occurred in northern Africa (Wasylikowa and van der Veen, 2004). It is thought that *C. lanatus* originated in the Kalahari region of Namibia and Botswana (Robinson and Decker-Walters, 1997; Ellul *et al.* 2007; Pitrat 2008).

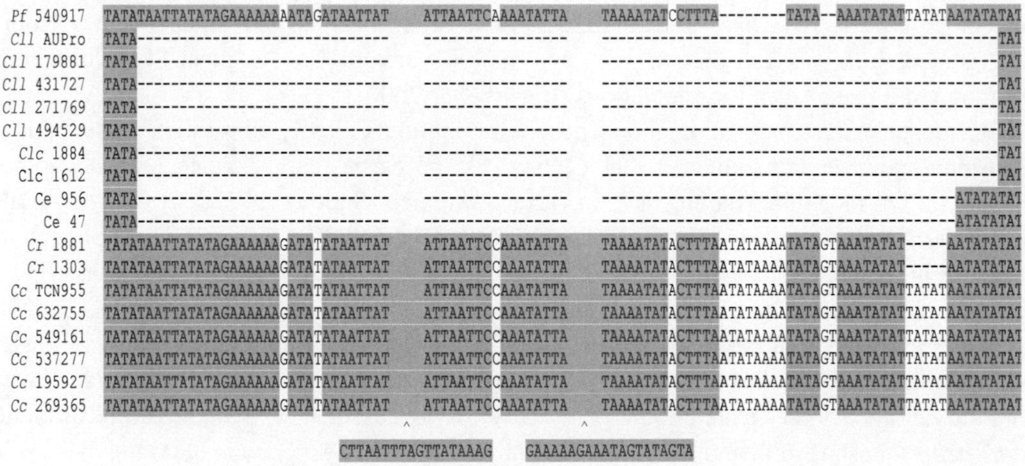

Figure 1. Sequence alignment of a 135 bp section of the *ycf*9 - *trn*G region of the chloroplast genome showing major deletions in *Citrullus lanatus* var. *lanatus* (*Cll* AU-Producer, PI's 179881, 431727, 271769, 494529); *C. lanatus* var. *citroides* (*Clc* 1884, 1612 from Namibia), and *C. ecirrhosus* (*Ce* 956 and 47 from Namibia) as compared to *C. rehmii* (*Cr* 1881 and 1303 from Namibia) and *C. colocynthis* (*Cc* TCN 955, PI's 632755, 549161, 537277, 195927, 269365). *Praecitrullus fistolosus* (*Pf* 540917) was used as outgroup.

One theory proposes that watermelon is derived from perennial *C. colocynthis*, endemic to Africa, another that watermelon was domesticated from putatively wild forms of *C. lanatus* var. *citroides* (Maynard, 2001). Chloroplast DNA investigations (Figure 1) of cultivated, egusi and citron-type watermelon collected from different areas of the African continent indicated divergence of cultivated and egusi into a lineage separate from the citron-type watermelon and evolution from a common ancestor, possibly *C ecirrhosus*, from Namibia (Dane *et al.* 2004; Dane and Lang 2004; Dane and Liu 2007).

Severe bottlenecks in the genetic background of cultivated watermelon have been reported based on DNA marker analysis of genetic similarities (Levi *et al.* 2001; Dane and Liu 2007). Watermelon seeds have also been recovered from Roman and Islamic deposits at Quseir al-Qadim, a port located on the Red Sea coast of Egypt. Seeds from the Roman deposits, dating back to 1st to 3rd CENT. CE, were of smaller size than seeds of Islamic deposits dating back to the 11 and 15th CENT. CE, suggesting watermelon consumption of sweet fruit flesh in Roman times and seeds in the Islamic period (Cox and van der Veen, 2008). Depictions of sweet watermelons and citrons have been found in the *Tacuinum Sanitatis*, based on a Latin translation of an 11 century Arabic manuscript *Taqwim al-Sihha bi al-Ashab al-Sitta*. Large oval, light and dark green striped fruits with red flesh and spherical, labeled as *Melones dulces,* and dark green fruits labeled as *Melones insipidi,* which was thought to represent cultivated citron watermelons also known as preserving melon (Paris *et al.* 2009). Watermelon seed remains from the Middle Ages were excavated from two sites in Hungary, Debrecen (13th CENT.) and Budapest (15th CENT.). Seed morphological analysis and SSR analysis of aDNA showed high similarities to red-fleshed Hungarian watermelon cultivars (Gyulai *et al.* 2008; Tóth *et al.* 2007).

C. colocynthis, the colocynth or bitter apple, has been cultivated in the Mediteranean region since 6000 BP (Zohary and Hopf 2000). Its small seeds appear in several Egyptian, Libyan and Near Eastern sites. On the island of Cyprus, the raising of colocynth has been a source of revenue since the 14th CENT. It is a drought-tolerant species which can survive arid environments by maintaining its water content without the wilting of leaves or desiccation under severe drought conditions. Sequence information from several polymorphic intergenic cpDNA regions and a large intron at the transit sequence of G3pdh, showed geographic structure among *C. colocynthis* accessions from many different regions. Divergent lineages of *C. colocynthis* from tropical Asia and Africa could be traced to migration routes from the Middle East (Dane *et al.* 2007).

Genus *Cucumis*

Cucumis sativus (2n = 2x = 14), the cucumber, which ranks among the top 10 vegetables in world production, is of Asian origin and has been domesticated since 3500 BP. India is considered the center of origin and domestication, where two botanical varieties *C. sativus* var. *sativus* L. (cultivated) and the feral form var. *hardwickii* (R.) coexist. The species has a narrow genetic base within domesticated market types (Staub *et al.* 2008). Chloroplast and nuclear short sequence repeat sequence information has been used to study phylogenetic relationships of the genus *Cucumis*. Three major species clusters, a group of African *Cucumis* species (2x = 24), a group composed of *C. melo* (2n = 2x = 24) accessions, and a group containing *C. sativus* and *C. hystrix* Chakr. (2n = 2x= 24) were identified, suggesting that *C. sativus* and *C. hystrix* share a common ancestral lineage (Chung *et al.* 2006). Analysis of DNA sequences from chloroplast gene intron and spacer regions and nuclear internal transcribed space (ITS) sequences from *Cucumis* and related genera indicated that cucumber and *C. hystrix* are nested among Australian, Malaysian and Western Indian species. *Cucumis melo* is sister to this Australian/Asian clade rather than being close to African *Cucumis* species as previously thought (Renner *et al.* 2007), which appear to have evolved more recently. Most of the species in the *Cucumis* clade are moneocious perennials, which may be the ancestral conditions from which the annual habit and dioecy appear to have evolved several times (Renner *et al.* 2007). The origin of *C. melo* remains unknown (Renner *et al.* 2007).

The first *Cucumis* remains were excavated from the Hoabinhian stratum in Thailand dating to approximately 10,000 to 6000 BCE (Chang, 1970), including cucumber type at least 7180 $\pm$ 360 BCE as analyzed by C^{14} of bamboo charcoal (Gorman, 1969). Cucumber is known from descriptions in Iraq at about 2600 BP and in the Mediterranean area at 2200 BP. Combining archeological records of seeds, iconography and analysis of available texts, it is thought that *C. melo* was domesticated probably around 5000 BP and was cultivated in China at 5000 BP, India 4000 BP, Egypt at 3500 BP (Pitrat *et al.* 1999; Pitrat 2008). While melons were an established crop in Egypt, there are no records indicating the existence of sweet melons in Greek and Romans times. Apparently sweet melons were first selected during the Middle Ages, arriving in Europe from Middle Asia and Asia Minor during the 14th and 15th century (Janick and Paull 2008). The oriental pickling melon (*C. melo* ssp. *melo* var. *conomon*) is considered the most ancient form of melon domesticated in China. Melon could have been domesticated for its seeds rich in proteins and lipids of good nutritional values since in most cases wild melon is bitter due to the presence of cucurbitacins. Moreover the flesh is very thin and domestication resulted in the development of a fruit mesocarp (Pitrat 2008). Molecular analysis of aDNA from seed remains of melon (Szabó *et al.* 2005; 2008) recovered from 15th CENT. in Budapest, Hungary, indicated that the medieval melon was closely related to a registered Hungarian cultivar selected from an old Hungarian landrace, indicating that melons introduced into Europe in the Middle Ages are still in cultivation today (Janick and Paull 2008).

Genus *Lagenaria*

The bottlegourd (*Lagenaria siceraria*) has been grown worldwide for thousands of years for its strong, hard-shelled and buoyant fruits priced as containers, music instruments and fishing floats (Heiser, 1979). It is the only cultivated plant present in both Early Holocene America and Asia. Morphological and genetic differences between present-day African and Asian bottle gourd cultivars can be used to divide the species into 2 subspecies, *L. siceraria* spp *siceraria* and *L. siceraria* spp *asiatica*, suggesting an ancient eastward diffusion of the species out of Africa. The recent integration of archaeological and genetic research on bottle gourd has clarified our understanding of its initial domestication, its late Pleistocene radiation from Asia across Beringia into the Americas (Erickson *et al.* 2005). Chloroplast DNA analysis using 2 intergenic regions resulted in 3 diagnostic polymorphisms, which could be used to distinguish Asian from African landraces of bottle gourd with DNA extracted from archaeological rind fragments. The archaeological rind fragments were identical to modern Asian bottle gourd landraces, indicating that the domesticated bottle gourds were carried to the New World during the late Pleistocene from Asia, not from Africa. A study of ancient chloroplast DNA extracted from a New World plant remains, dating more than 6000 years BP, documented the dispersal of morphologically domesticated bottle gourd in the New World from Asia (Erickson *et al.* 2000). Also this research supported the domestication of bottle gourd in the New World as early as 10,000 BP (Erickson *et al.* 2005) and suggests that *L. siceracia* was independently domesticated at least twice. The bottle gourd has been recovered from archaeological contexts in China and Japan dating to ca 8000 to 9000 BP. whereas in Africa, despite decades of high-quality archaeobotanical research, the earliest recorded report of recovery of a bottle gourd is from the 12th Dynasty tomb at Thebes dating to 4000 BP. Thus bottle gourd might have been domesticated in Asia first and 4000 years later in Africa.

CONCLUSIONS

The Cucurbitaceae family contains some of the first plant species to be domesticated for human use, providing food, medicine, and a wide variety of utensils and even instruments. One of the earliest domesticated plants, *L. siceracia* is pivotal to issues concerning the mechanism and cultural parameters of the earliest human transitions to plant domestication (Clarke *et al.* 2006). In this economically important family difficulties with human transport of seeds between continents for at least 10,000 years have led to the ambiguous geography of the closest relatives of watermelon, and cucumber as well as other less cultivated cucurbits such as loofah (*Luffa acutangula*) and bitter gourd (*Momordica charantia*). Worldwide sampling and chloroplast DNA analyses of many Cucurbitacea species revealed an Asian origin in the Late Cretaceous, specifically the region north of the Tethys in Asia, followed by repeated spread of lineages into the African, American and Australian continents via transoceanic long-distance dispersal (Schaefer and Renner 2009). North American cucurbits stem from at least seven range expansions, while cucurbit species in Europe are probably the result of recent introductions or casual escapes from cultivated species. With genome sequencing of the first cucurbit species (*C. sativus)* and the development of extensive and detailed genetic and physical maps (Ren *et al.* 2009; Ezura and Fukino 2009; Gonzalez *et al.* 2010; Guo *et al.* 2010) of domesticated species, genes and traits can be linked and genes selected during domestication can be characterized and serve as excellent resources for the study of the complex domestication and cultivation history of cucurbit crops.

REFERENCES

Alverson AJ, Wei XX, Rice DW, Stern DB, Barry K, Palmer JD (2010) Insights into the evolution of mitochondrial genome size from complete sequences of *Citrullus lanatus* and *Cucurbita pepo* (Cucurbitaceae). *Mol Biol Evol* 27: 1436-1448.

Arumuganathan K and Earle E (1991) Nuclear DNA content of some important plant species. *Plant Mol Biol Rep* 9: 208-218.

Chang KC (1970) The beginnings of agriculture in the Far East. *Amer Antiquity* 35: 175-184.

Chung S-M, Staub JE, Chen J-F (2006) Molecular phylogeny of *Cucumis* species as revealed by consensus chloroplast SSR marker length and sequence variation *Genome* 49: 219-229.

Clarke AC, Burtenshaw MK, McLenachan PA, Erickson DL, Penny D (2006) Reconstructing the origins and dispersal of the Polynesian bottle gourd (*Lagenaria siceraria*). *Molecular Biology Evolution* 23:893-900.

Cox A and van der Veen M (2008) Changing foodways: watermelon (*Citrullus lanatus*) consumption in Roman and Islamic Quseir al-Qadim, Egypt. *Veget Hist Archaeobot* 17:181-189.

Dane F and Lang P (2004) Sequence variation at cpDNA regions of watermelon and related species: implications for the evolution of *Citrullus* haplotypes. *Amer J Bot* 91: 1922-1929.

Dane F, Lang P, Backtiyarova R (2004) Comparative analysis of chloroplast variability in wild and cultivated *Citrullus* species. *Theor Appl Genet* 108: 958-966.

Dane F and Liu J (2007) Diversity and origin of cultivated and citron type watermelon (*Citrullus lanatus*). *Genet Resour Crop Evol* 54:1255-1265.

Dane F, Liu J, Zhang C (2007) Phylogeography of the bitter apple, *Citrullus colocynthis*. *Genet Resour Crop Evol* 54: 327-336.

Decker-Walters DS, Staub JE, Chung S-M, Nakata E, Quemada HD (2002) Diversity in free-living populations of *Cucurbita pepo* (Cucurbitaceae) as assessed by random amplified polymorphic DNA. *Econ Bot* 27: 19-28.

Decker-Walters and Walters TW (2000). Squash. In Kiple KF, Ornelas KC (Eds) *The Cambridge world history of food*. Cambridge Univ Press, Cambridge UK. pp 335-351.

Ellul P, Lelivelt C, Naval MM, Noguera FJ, Sanchez S, Atares A, Moreno V, Corella P, Dirks R (2007) Watermelon In: Biotechnology in agriculture and forestry 60. Transgenic crops V. Pua EC, Davey MR (Eds). Springer-Verlag, Berlin pp129-162.

Emshwiller E (2006) Genetic documentation of plant domestication. In: *Documenting domestication. New genetic and archaeological paradigms*. Zeder M, Bradley DG, Emshwiller E, Smith BD (Eds) Univ CA Press, Berkeley CA pp. 99-122.

Erickson DL, Smith BD, Clarke AC, Sanweiss DH, Tuross N (2005) An Asian origin for a 10,000-year-old domesticated plant in the Americas. *Proc. Natl. Acad. Sci. USA* 102: 18315-18320.

Ezura H and Fukino N (2009) Research tools for functional genomics in melon (*Cucumis melo*): Current status and prospects. *Plant Biotech* 26: 359-368.

Ferriol M, Pico B, de Cordova PF, Nuez F (2004) Molecular diversity of a germplasm collection of squash (*Cucurbita moschata*) determined by SRAP and AFLP markers. *Crop Sci* 44: 653-664.

Janick J and Paull RE (2008) *Encyclopedia of fruit and nuts*. CAB International. London, UK.

Jeffrey C (2001) Ccurbitaceae In: Hanelt P *et al.* (eds) *Mansfeld's encyclopedia of agricultural and horticultural crops*. Springer, Berlin.

Jeffrey C (2005) A new system of Cucurbitaceae. *Bot Zhurn* 90: 332-335.

Gonzalez VM, Garcia-Mas J, Arus P, Puigdomenech P (2010) Generation of a BAC-based physical map of the melon genome. *BMC Genomics* 11:339.

Gorman CF (1969) Hoabinhian: a people-toll complex with early plant associations in southeast Asia. *Science* 163: 671-673.

Guo S, Zheng Y, Young J-G, Liu S, Zhang Z, Crasta OR, Sobral BW, Xu Y, Huang S, Fei Z (2010) Transcriptome sequencing and comparative analysis of cucumber flowers with different sex types. *BMC Genomics* 11: 384.

Gyulai G, Waters L, Dane F (2008) Ancient *Citrullus* DNA-unlocking domestication events. Fulbright Grant AY 2005-2006. Fulbright Year Book 2008, Budapest, pp 1-13.

Heiser CB (1979) *The gourd book*. Univ of Oklahoma Press, Norman, OK.

Hepper FN (1990) *Pharaoh's flowers. The botanical treasures of Tutankhamun*. Royal Botanic Gardens, Kew, London.

Huang S, LiR, Zhang Z, Li L, *et al.* (2009) The genome of the cucumber, *Cucumis sativus* L. *Nature Genetics* 41: 1275-1281.

Katzir N, Tadmor Y, Tzuri G, Leshzeshen E, Mozes-Daube N, Danin-PolegY, Paris HS (2000) Further ISSR and preliminary SSR analysis of relationships among accessions of *Cucurbita pepo*. *Acta Horticulturae* 510: 433-439.

Kocyan A, Zhang L-B, Schaefer H, Renner SS (2007) A multi-locus chloroplast phylogeny for the Cucurbitaceae and its implications for character evolution and classification. *Mol Phylogenet Evol* 44: 553-577.

Levi A, Thomas CE, Wehner TC, Zhang X (2001). Low genetic diversity indicates the need to broaden the genetic base of cultivated watermelon. *HortScience* 36: 1096-1101.

Maynard DN (2001) *An introduction of the watermelon*. ASHS Press, Alexandria, VA.

Paris HS, Daunay MC, Janick J (2009) The Cucurbitaceae and Solanaceae illustrated in medieval manuscripts known as the *Tacuinum Sanitatis*. *Annals of Botany* 103:1187-1205.

Pitrat M (2008) Melon. In: *Handbook of plant breeding* Vol I Part IV. Springer NY pp 283-315.

Pitrat M, Chauvet M, Foury C (1999) Diversity, history and production of cultivated cucurbits. *Acta Horticulturae* 492:22-28.

Ren Y, Zhang Z, Liu J, Staub JE, Han Y, Cheng Z, Li X, Lu J, Miao H, Kang H, Xie B, Gu X, Wang X,Du Y, Jin W, Huang S (2009) An integrated genetic and cytogenetic map of the cucumber genome *Plos One* 4: e5795.

Renner S, Schaefer H, Kocyan A (2007) Phylogenetics of *Cucumis* (Cucurbitacea): Cucumber *(C. sativus)* belongs in an Asian/Australian clade far from melon (*C. melo*). *BMC Evolutionary Biology* 7: 58.

Robinson RW and Decker-Walters DS (1997) Cucurbits. CAB International, New York.

Sanjur OI, Piperno DR, Andres TC, Wessel-Beaver L (2002) Phylogenetic relationships among domesticated and wild species of *Cucurbita* (Cucurbitaceae) inferred from a mitochondrial gene: implications for crop plant evolution and areas of origin. *Proc Nat Acad Sci USA* 99: 535-540.

Schaefer H, Heibl C, Renner SS (2009). Gourds afloat: a dated phylogeny reveals an Asian origin of the gourd family (Cucurbitaceae) and numerous oversea dispersal events. *Proc R Soc B* 276: 843-851.

Smith BD (1997) The initial domestication of *Cucurbita pepo* in the Americas 10,000 years ago. *Science* 276:932-934.

Smith BD (2005) Reassessing Coxcatlan Cave and the early history of domesticated plants in Mesoamerica. *Proc Natl Acad Sci* USA 102:9438-9445.

Smith RD (2006) Seed size increase as a marker of domestication in squash (*Cucurbita pepo*). Documenting domestication. *New genetic and archaeological paradigms*. Zeder MA et al. Eds. Univ CA Press, Berkely CA. pp 25-31.

Smith BD and Yarnell RA (2009) Initial formation of an indigenous crop complex in eastern North America at 3800 B.P. *Proc Natl Acad Sci USA* 106: 6561-6566.

Staub JE, Robbins MD, Wehner TC (2008) Cucumber In: Prohens J, Nuez F (eds) *Vegetables* I. Springer New York pp 241-282.

Szabó Z, Gyulai G, Tóth Z, Heszky L (2008) Morphological and molecular diversity of 47 melon (*Cucumis melo*) cultivars compared to an extinct landrace excavated from the 15th century. *Cucurbitaceae 2008 Proceedings* (Pitrat M ed). pp 313-321.

Szabó Z, Gyulai G, Humphreys M, Horváth L, Bittsánszky A, Lágler R, Heszky L (2005) Genetic variation of melon (*C. melo*) compared to an extinct landrace from the Middle Ages (Hungary) I. rDNA, SSR and SNP analysis of 47 cultivars. *Euphytica* 146: 87-94.

Tóth Z, Gyulai G, Szabó Z, Horváth L, Heszky L (2007) Watermelon (*Citrullus l. lanatus*) production in Hungary from the Middle Ages (13th century). *Hungarian Agricultural Research* 4:14-19.

Wasylikowa K and van der Veen M (2004) An archaeobotanical contribution to the history of watermelon, *Citrullus lanatus* (Thunb.) Matsum. & Nakai (syn *C. vulgaris* Schrad.). *Veget Hist Archaeobot* 13: 213-217.

Zeist van W (1983) Fruits in foundation deposits of two temples. *J Archaeol Sci* 10: 351-354.

Zohary D and Hopf M (2000) domestication of plants in the old world. Oxford University Press, Oxford UK.

III. Additional Sources for Archaeogenetics

In: Plant Archaeogenetics
Editor: Gábor Gyulai, pp. 117-127

Chapter 10

CONSERVATION GENETICS (1710 - 2010) – CLONING OF LIVING FOSSILS: MICROPROPAGATION OF THE OLDEST HUNGARIAN BLACK LOCUST TREE (*Robinia pseudoacacia*) PLANTED IN 1710 (BÁBOLNA, HUNGARY)

Gábor Gyulai[1,2,3,*], Réka Láposi[1,4], Heinz Rennenberg[2], Anikó Veres[1], Cornelia Herschbach[2], Gyula Fábián[4] and Luther Waters Jr.[5]

[1]Institute of Genetics and Biotechnology, St. Stephanus University,
Gödöllő 2103 Hungary
[2]Institute for Forest Botany and Tree Physiology, Freiburg University,
Freiburg 79085 Germany
[3]Deutscher Akademischer Austauschdienst (DAAD),
Bonn 53175 Germany
[4]Institute of Environmental Sciences, Károly Róbert College,
Gyöngyös 3200 Hungary
[5]Department of Horticulture, College of Agriculture,
Auburn University, Alabama 36849 USA

ABSTRACT

Delicate species, old varieties and heirlooms are threatened by extinction due to changes in environmental conditions, over cultivation, and falling below the minimum viable population. Conservation genetics provides effective tools of micropropagation to produce a large number of clones. As the clones develop from somatic tissues or organs, the genome (DNA content) remains identical in each clone (Gyulai *et al.* 2003). In the

---

[*] E-mail: gyulai.gabor@mkk.szie.hu

case of the 300-year-old *Robinia pseudoacaia cv. 'Bábolna'*, micropropagation was successfully used for clonal propagation resulting in identical clones. After rooting, clones will be planted in an archaeogenetical garden of botany in Gödöllő (Hungary).

INTRODUCTION

Black locust (*Robinia pseudoacacia*) (family *Fabaceae*) is a multipurpose, drought tolerant, open pollinated, nitrogen fixing, ornamentally attractive tree. Although it is native to the southeastern United States, it has been widely planted and naturalized in the northern hemisphere for its extremely hard and rot resistant wood.

Fossils of *Robinia*-like trees (*Robinia zirkelii*) (*syn.: Cercidoxylon / Robinoxylon z.*; *Robinia / Robinoxylon brewerii*; and *Paleo-Robinoxylon z.*) were found in several sites in North America, including the oldest *Robinia zirkelii* fossils from the Late Eocene era (Chadronian; 38 - 33.9 Ma, million year BP) in Nebraska, USA (Wheeler and Landon 1992). *Robinia* became extinct in Europe, but fossils were found in Miocene flora (Middle Badenian; 14.3 to 3.8 Ma) (Böhme *et al.* 2007).

The oldest black locust in Europe was planted in 1601 (transplanted? in 1635) by Jean Robin who introduced it to Europe (see *Robinia*, given the name by Linnaeus), and it is still growing in Paris (Square René Viviani, Paris, France).

In Hungary, the oldest black locust was planted in 1710, and is still growing in Bábolna (*Robinia pseudoacacia cv. Bábolna*) (Figure 1). Some decades later, Sámuel Tessedik (1742-1820) developed thousands of acres of black locust plantations in Hungary. It extended from 276,000 ha, 18.2 % (Keresztesi 1983; Walkovszky 1998) to 23 % (410 000 ha) of the total forested land (1.46 Mha) of Hungary (Rédei and Veperdi 2009).

The third oldest European black locust, the *'Old Lion'* planted in 1762, still grows in the Royal Botanical Garden, Kew (England).

Furthermore, *Robinia* species were introduced to Europe: *Robinia hispida* in 1743, and *Robinia viscosa* in 1797 (Földes 1903; Peabody 1982).

The aim of this Chapter is to describe the use of micropropagation for cloning the oldest Hungarian black locust tree in aseptic bud culture to maintain its gene pool. Genetic analysis of the ancient clones '*Bábolna*' was also carried out with comparison to current black locust clones and related species to track molecular changes that have occurred during the past 300 years.

MATERIALS AND METHODS

Plant Materials

The black locust tree *Robinia pseudoacaia cv. 'Bábolna'* (planted in 1710) was compared to *R. pseudoacacia 'Paris'* (planted in 1602 - 1635) (Square René Viviani, Paris, Fr) (sampled by DÁ Daniek, 2010), current clones grown in Gyöngyös and Tass-Puszta (Atkár, H); and thornless clone of *R. pseudoacacia inermis* sampled from the Arboretum Gödöllő, Hungary (H).

Figure 1. Route of the oldest (from 1710) Hungarian black locust (*Robinia pseudoacacia cv. 'Bábolna'*) from Northern America to Hungary, Europe, city Bábolna.

Figure 2. The oldest Hungarian black locust tree (*Robinia pseudoacacia cv. 'Bábolna'*) planted in 1710. *(a)* The memorial plate, translation: *The black locust seeds were taken from North America in 1710... .* *(b)* The tree in the winter (Photo by Réka Láposi) and *(c)* in the spring (Photo by Zsigmond G Gyulai). *(d)* blooming branches, *(e)* thorn, *(f)* the *polyphylla* type leaf, *(g)* the small two seeded pod, and *(h)* the micropropagated clones developed from bud culture *in vitro*.

Further hybrids and related species were also included in the study: *Robinia x ambigua* (Erdőtelek, H); *R. viscosa,* clammy locust, Gyöngyös; *R. neomexicana* Vácrátót; *Amorpha fruticosa*, desert false indigo (Tass, H); *Gleditsia triacanthos*, honey locust (Erdőtelek, H); *Sophora japonica*, Japanese pagoda tree (Erdőtelek, H); *Caragana arborescens*, Siberian peashrub (Erdőtelek H); *Gymnocladus dioicus*, Kentucky coffee tree (Erdőtelek, H) and *Laburnum anagyroides*, golden chain (Erdőtelek, H).

Aseptic Bud Culture

Shoot apical buds were dissected and processed for aseptic shoot culture following the general tissue culture protocols (Gyulai *et al.* 2006). Buds were cleaned, washed with detergent (3 min) and rinsed three times with distilled water (3 min), followed by surface sterilization with ethanol (70% v/v) for 1 min and a commercial bleaching agent (8% NaOCl w/v) for 1 min; followed by three rinses with sterile distilled water, and incubated in aseptic tissue culture medium F6 (Gyulai *et al.* 2003) supplemented with 0.1 mg/L 2,4-dichlorophenoxyacetic acid (2,4-D) and kinetin, respectively. In a comparative experiment IBA (indole-3-butyric acid) was used (1 mg/L) suggested by Barghchi (1987). Clones, sprouted from aseptic buds in three weeks (Figure 2) were further micropropagated with nodal segments. After the subsequent incubation time, clones were rooted on hormone-free medium, transplanted to pots (Shu *et. al* 2003) and grown in green houses.

Molecular Analyses. DNA Extraction

Young fresh leaves (0.1 g) were ground in an aseptic mortar with liquid nitrogen. DNA was extracted by the CTAB (cethyltrimethylammonium bromide) method according to the modification of Gyulai *et al.* (2006) and Bisztray *et al.* (2004) followed by an RNase-A treatment (Sigma, R-4875) for 30 min at 37° C in each case.

The quality and quantity of extracted DNA was measured (2 μl) by a NanoDrop ND-1000 UV-Vis spectrophotometer (NanoDrop Technologies, Delaware, USA – BioScience, Budapest, Hungary). DNA samples were adjusted to a concentration of 30 ng/μl with ddH$_2$O and subjected to PCR amplification.

PCR Amplification

Hot Start PCR was combined with Touchdown PCR using AmpliTaq Gold^{TM} Polymerase. Reactions were carried out in a total volume of 25 μl containing genomic DNA of 30-50 ng, 1 x PCR buffer (2.5 mM MgCl$_2$), dNTPs (200 μM each), 20 pmol of each primer and 1.0 U of *Taq* polymerase. Touchdown PCR was performed by decreasing the annealing temperature by 1.0 °C / per cycle with each of the initial 12 cycles (PE 9700, Applied Biosystems), followed by a 'touchdown' annealing temperature for the remaining 25 cycles at 56 °C for 30 s with a final cycle of 72 °C for 10 min and held at 4 °C. A minimum of three independent DNA preparations of each sample was used. Amplifications were assayed by agarose (1.8 %, SeaKem LE, FMC) gel electrophoresis, stained with ethidium bromide (0.5 ng/µl) after running at 80 V in 0.5 x TBE buffer. Each successful reaction with scorable bands was repeated at least twice. Transilluminated gels were analyzed by the ChemilImager v 5.5 computer program (Alpha Innotech Co.). A negative control which contained all the necessary PCR components except template DNA, was included in the PCR runs.

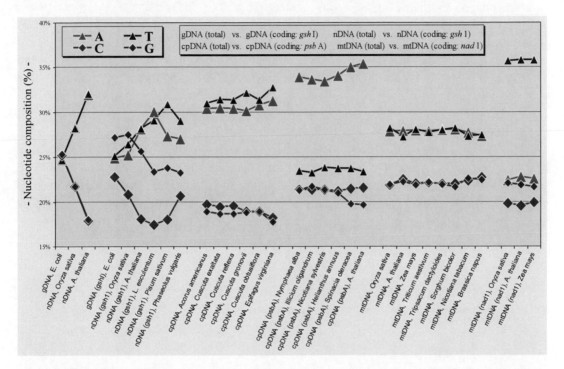

Figure 3. Nucleotide compositions (A – adenine; T – thymine, C – cytosine, G – guanine) of coding and non-coding parts of genomes comparing DNA sequences of genomic (total, gDNA) and coding, nuclear (nDNA) and organelle DNA sequences (cpDNA and mtDNA). The following sequences were compared (NCBI accession #): gDNA *E. coli* (NC_003070); nDNA *Oryza sativa* (NC_008402.1); nDNA *A. thaliana* (NC_003076.5). // gDNA (*gsh*I) *E. coli* (X03954); nDNA (gsh1) *Oryza sativa* (AJ508916); nDNA (*gsh*1) *A. thaliana* (Y09944); nDNA (*gsh*1) *L. esculentum* (AF017983); nDNA (*gsh*1) *Pisum sativum* (AF128455); nDNA (*gsh*1) *Phaseolus vulgaris* (AF128454). // cpDNA *Acorus americanus* (NC_010093.1); cpDNA *Cuscuta exaltata* (NC_009963.1); cpDNA *Cuscuta reflexa* (NC_009766.1); cpDNA *Cuscuta gronovii* (NC_009765.1); cpDNA *Cuscuta obtusiflora* (NC_009949.1); cpDNA *Epifagus virginiana* (NC_001568.1). // cpDNA (*psb*A) *Nymphaea alba* (NC_006050); cpDNA (*psb*A) *Illicium oligandrum* (EF380354); cpDNA (*psb*A) *Nicotiana sylvestris* (AB237912); cpDNA (*psb*A) *Helianthus annuus* (DQ383815); cpDNA (*psb*A) *Spinacia oleracea* (AJ400848); cpDNA (psbA) *A. thaliana* (AP000423). // mtDNA *Oryza sativa* (NC_007886.1); mtDNA *A. thaliana* (NC001284.2); mtDNA *Zea mays* (NC_007982.1); mtDNA *Triticum aestivum* (NC_007579.1); mtDNA *Tripsacum dactyloides* (NC_008362.1); mtDNA *Sorghum bicolor* (NC_008360.1); mtDNA *Nicotiana tabacum* (NC_006581.1); mtDNA *Brassica napus* (NC_008285). // mtDNA (*nad*1) *Oryza sativa* (NC_007886); mtDNA (*nad*1) *A. thaliana* (NC_001284); mtDNA (*nad*1) *Zea mays* (NC_007982).

PCR Primers

Two microsatellites of cpDNA at loci *scu*7 and *ccmp*6, and a nuclear ITS (*i*nternal *t*ranscribed *s*pacer of rDNA; ITS1-5.8S-ITS2) primer pairs were used. For *scu*7 5'-*cct aac ttg aaa cga aag gac tgc*-3' and 5'-*ccg aag agg aat atg ggt ttg ag*- 3' (Scott *et al.* 2000); for *ccmp*6 5'-*gtt tca ttc ggc tcc ttt at*-3' and 5'-*cga tgc ata tgt aga aag cc*- 3' (Weising and Gardner 1999); and for ITS 5'-*gaa cga ccc ggg aac acg tt*-3' and 5'-*gct gcg ttc ttc atc gat gc*-3' (White *et al.* 1990) were used. The cpDNA analyses (*ccmp*6) required improved PCR cycling (Mansour *et al.* 2009) due to the special nucleotide compositions of organelle DNAs (cpDNA and mtDNA) compared to nuclear nDNA at both coding and non-coding loci (Figure 3).

ALF Analysis

Microsatellites of cpDNA and ITS fragments of nuclear DNA were forwarded for ALF analysis using ALF ExpressII (Amersham - AP-Hungary, Budapest). One strand of each of the SSR primer pairs was labeled with Cy5 dye. PCR and ALF analysis were carried out by the protocol of Röder *et al.* (1998), Huang *et al.* (2002) and Gyulai *et al.* (2006) (Figure 4a,b).

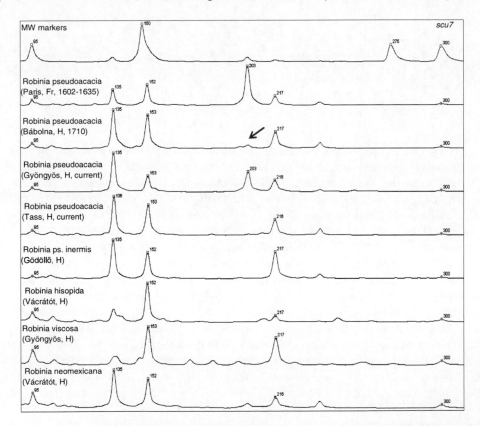

Figure 4a. ALF-SSR patterns of *Robinia* species and clones at the *scu7* locus of cpDNAs with a missing fragment (203 bp) in the *Bábolna* clone planted in 1710 (Hungary) compared to clone of *Paris* planted in 1602, and current clone *Gyöngyös* and *Tass*; and other *Robinia* species (MW size marker bps 95, 150, 275 and 300).

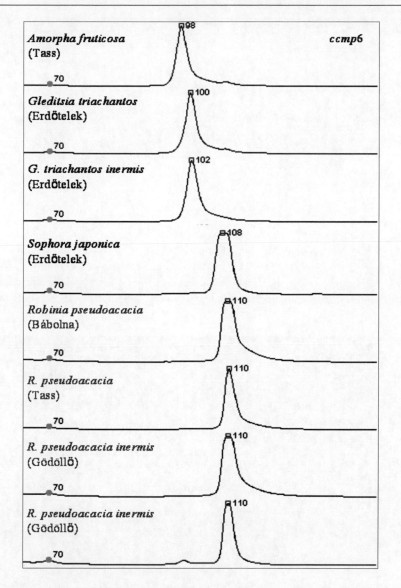

Figure 4b. ALF-SSR patterns of the species and clones investigated at the *ccmp*6 locus of cpDNAs with increased fragment sizes (98-100 bp) from *Amorpha* to *Robinia* including clone '*Bábolna*' planted in 1710 Bábolna, Hungary (size marker 70 nt).

CONCLUSION

Natural clones of the world's oldest trees indicate the importance of clonal propagation such as in the cases of an 11,700-year-old creosote bush (*Larrea tridentata*), the 'King Clone' (Mojave Desert, Lucerne Valley, California, USA) and the 9500-year-old Norway spruce (*Picea abies*) the 'Old Tjikko' (Sweden).

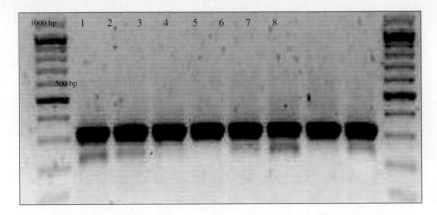

Figure 5a. ITS. ITS analysis of *Robinia* samples and related species. *(1) R. pseudoacacia Paris*, Fr, (1602-1635), *(2) R. pseudoacacia Bábolna*, Hungary (1710), *(3)* current trees of *R. pseudoacacia* Gyöngyös H, *(4) R. pseudoacacia* Tass H, *(5) R. pseudoacacia inermis* Gödöllő H, *(6) R. hispida* Vácrátót H, *(7) R. viscosa* Gyöngyös H, and *(8) R. neomexicana* Vácrátót H (in the same order as in Figure 4a) (MW size markers are indicated; samples were loaded by Katica Veres and Krisztina Simon).

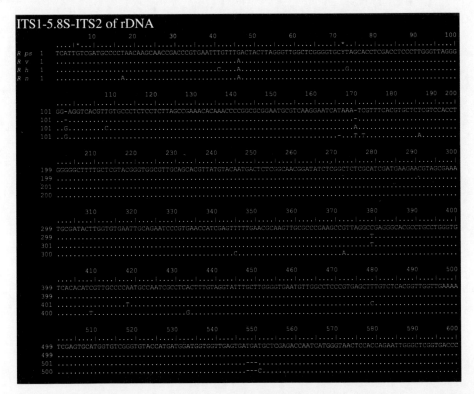

Figure 5b. ITS. Consensus sequence alignments of data bank accessions (NCBI - Altschul *et al.* 1997 and BioEdit - Hall 1999) of ITS1-5.8S-ITS2 sequences of species *Robinia*. *R ps - Robinia pseudoacacia* (AF174637.1); *R v - R. viscosa* (AF398821); *R h - R. hispida* (AF398819.1); *R n - R. neomexicana* (AF537350.1) (SNPs are indicated). * indicates clonal SNPs at the 6th nt (G / -) and 72th nt (Y [T or C]T / T) among seven accessions of *R. pseudoacacia* (Accession #: EF494737.1; AF176391.1; AF176390.1, AF176389.1; AF174637.1; DQ006010.1; AF398820.1).

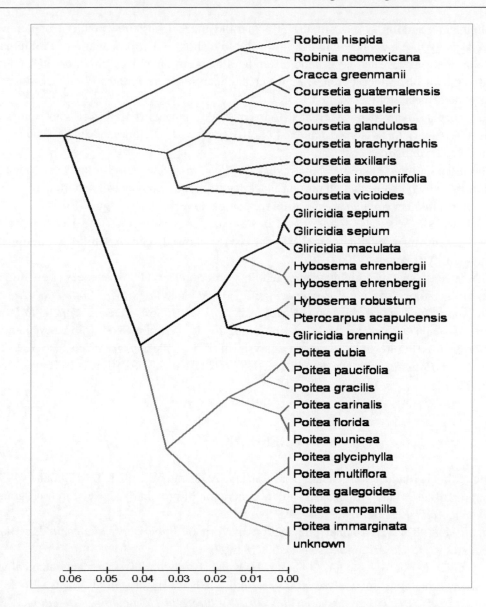

Figure 5c. Molecular dendrogram (MEGA4, Tamura *et al.* 2007) of *Robinia* and related species (*Fabaceae*) based on sequence homology at the loci ITS1-5.8S-ITS2. Scale genetic distance. (Accession #: AF398772.1; AF398782.1; AF398781.1; AF398777.1; AF398775.1; AF398803.1; AF398802.1; AF398795.1; AF398793.1; AF398788.1; AF398791.1; AF398769.1; AF398815.1; AF398812.1; AF398814.1; AF398770.1; AF398768.1; AF398808.1; AF189055.1; AF398850.1; AF542457.1; GQ996251.1; AF398840.1; AF398839.1; AF398819.1; AF537351.1; AF398849.1; AF542461.1; GQ996252.1.)

To conclude, clonal propagation of a 300-year-old black locust tree grown in city Bábolna (Hungary) was applied successfully for maintaining a 300-year-old black locust genome. The method might also be applied to other 'living fossils' such as the 4789-year-old Basin Bristlecone Pine (*Pinus longaeva*) the '*Methuselah*' (Wheeler Peak, eastern Nevada, USA), the 4-5000-year-old yew (*Taxus baccata*) (Llangernyw, Conwy County Borough, North Wales, UK), and even to resurrect *Prometheus*, formerly the oldest organism (4862-year-old *Pinus longaeva*), from cells or deep buds as it was cut down in 1964 as a research mistake.

Clonal propagation *in vitro* of the 300-year-old black locust tree in this Chapter was achieved using aseptic buds cultures followed by shoot micropropagation. Preliminary molecular analyses revealed no genetic differences between the clones of '*Bábolna*' and the currently grown black locust clones, however the analyses were found to be discriminative among related species at loci studied.

Polymorphic locus at *scu*7 revealed four cpDNA fragments (Figure 4a) with a missing fragment (203 bp) form clone '*Bábolna*' (1710) compared to clones of *Paris* (1602) and current clone *Gyöngyös*.

The monomorphic locus at *ccmp*6 revealed a *Robinia*-species specific cpDNA fragment (110 bp) in *Bábolna* clone and all current black locust clones, compared to shorter fragments (98 to 102 nt) in other *Robinia* species used for comparative analyses (Figure 4b).

Nuclear SSRs were also found useful in *Robinia* genotyping (Lian and Hogetsu 2002), however certain microsatellites like (AG)n (*Rops*15) showed high variability with somatic hypermutability (Lian *et al.* 2004).

The species specific, highly conserved locus of nuclear ITS sequences were highly monomorphic among species of *Robinia* (Figure 5a) and clade *Robinieae* (*Coursetia, Cracca, Genistidium, Gliricidia, Hebestigma, Hybosema, Lennea, Olneya, Peteria, Poitea, Robinia, Sphinctospermum*) (*Robinioid* clade) (Lavin *et al.* 2003) with rare SNPs (<u>S</u>ingle <u>N</u>ucleotide <u>P</u>olymorphism) (Figure 5b). Sequence analysis of ITS sequences *in silico* also revealed sequences of *Robinia* from Hancock Cave, USA (NCBI # AF176391.1; AF176390.1 and AF176389.1).

REFERENCES

Altschul SF, TL Madden, AA Schaffer, JH Zhang, Z Zhang, W Miller, DJ Lipmand (1997) Gapped BLAST and PSI-BLAST: a new generation of protein database search programs. *Nucleic Acids Research* 25: 3389-3402.

Barghchi M (1987) Mass clonal propagation in vitro of *Robinia pseudoacacia* L. (Black locust) cv. '*Jászkiséri*'. *Plant Science* 53: 183-189.

Bisztray GyD, R Bacsó, P Bodor, G Facsar, F Gyulai, I Velich (2004) Archaeobotanical and genetical methods to analyse 600-years-old seeds of horticultural plants. Book of Abstracts 5th IVCHB Symposium, *In Vitro Culture and Horticultural Breeding*,12-17. September 2004, Debrecen, Hungary, p. 212.

Böhme M, AA Bruch, A Selmeier (2007) The reconstruction of Early and Middle Miocene climate and vegetation in Southern Germany as determined from the fossil wood flora. Palaeogeography, *Palaeoclimatology, Palaeoecology* 253: 91 - 114.

Földes J (1903) Adalékok az akác ismeretéhez ('Further data about *Robinia*'). *Erdészeti Lapok* 1903/I: 63-65.

Gyulai G, Z Mester, J Kiss, Szemán L, Heszky L, A Idnurm (2003) Somaclone breeding of reed canarygrass (*Phalaris arundinacea* L). *Grass and Forage Sci* 58: 210-215.

Gyulai G, M Humphreys, R Lágler, Z Szabó, Z Tóth, A Bittsánszky, F Gyulai, L Heszky (2006) Seed remains of common millet from the 4th (Mongolia) and 15th (Hungary) centuries: AFLP, SSR and mtDNA sequence recoveries. *Seed Science Research* 16: 179-191.

Hall TA (1999) BioEdit: a user-friendly biological sequence alignment editor and analysis program for Windows 95/98/NT. *Nucelic Acids Symposium* Series No. 41: 95-98.

Huang XQ, A Börner, MS Röder, MW Ganal (2002) Assessing genetic diversity of wheat (*Triticum aestivum* L.) germplasm using microsatellite markers. *Theor Appl Genet* 105: 699-707.

Keresztesi B (1983) Breeding and cultivation of black locust (*Robinia pseudoacacia* L.) in. Hungary. *Forest Ecology and Management* 6: 217-244.

Lavin M, MF Wojciechowski, P Gasson, CE Hughes, E Wheeler (2003) Phylogeny of robinioid legumes (*Fabaceae*) revisited: *Coursetia* and *Gliricidia* recircumscribed, and a biogeographical appraisal of the Caribbean endemics. *Systematic Botany* 28: 387-409.

Lian C, T Hogetsu (2002) Development of microsatellite markers in black locust (Robinia pseudoacacia) using a dual-supression-PCR technique. *Molecular Ecology Notes* 2: 211-213.

Lian C, R Oishi, N Miyashita, T Hogetsu (2004) High somatic instability of a microsatellite locus in a clonal tree, *Robinia pseudoacacia*. *Theor Appl Genet* 108:836-841.

Mansour A, JA T da Silva, G Gyulai (2009) Assessment of Molecular (Dis)similarity: The Role of Multiple Sequence Alignment (MSA) Programs in Biological Research. *Genes, Genomes and Genomics* 3:23-30.

Peabody FJ (1982) A 350-Year-Old American Legume in Paris. *Castanea* 47: 99-104.

Rédei K, I Veperdi (2009) The role of black locust (*Robinia pseudoacacia* L.) in the establishment of short-rotation energy plantations in Hungary. *International Journal of Horticultural Science* 15: 41-44.

Röder MS, V Korzun, K Wendehake, J Plaschke, MH Tixier, P Leroy, MW Ganal (1998) A microsatellite map of wheat. *Genetics* 149: 2007-2023.

Scott KD, P Eggler, G Seaton, M Rosetto, EM Ablett, LS Lee, RJ Henry (2000) Analysis of SSRs derived form grape ESTs. *Theor Appl Genet* 100: 273-726.

Shu QY, GS Liu, DM Qi, CC Chu, J Li, HJ Li (2003) An effective method for axillary bud culture and RAPD analysis of cloned plants in tetraploid black locust. *Plant Cell Reports* 22:175-180.

Tamura K, Dudley J, Nei M, Kumar S (2007) MEGA4: Molecular Evolutionary Genetics Analysis (MEGA) software version 4.0. Mol *Biol Evol* 24: 1596-1599.

Walkovszky A (1998) Changes in phenology of the locust tree (*Robinia pseudoacacia* L.) in Hungary. *Int J Biometeorol* 41:155-160.

Weising K, RC Gardner (1999) A set of conserved PCR primers for the analysis of simple sequence repeat polymorphisms in chloroplast genomes of dicotyledonous angiosperms. *Genome* 1999(1): 9-19.

Wheeler EA and J Landon (1992) Late Eocene (Chadronian) dicotyledonous woods from Nebraska: evolutionary and ecological significance. *Review of Palaeobotany and Palynoiogy* 74: 267-282.

White TJ, T Bruns, S Lee, J Taylor (1990) Amplification and direct sequencing of fungal ribosomal RNA genes for phylogenetics. Chapter 38. In: *PCR Protocols: a Guide to Methods and Applications* (M Innis, D Gelfand, J Sninsky, T White, eds.). Academic Press, Orlando, Florida.

In: Plant Archaeogenetics
Editor: Gábor Gyulai, pp. 129-134

ISBN: 978-1-61122-644-7
© 2011 Nova Science Publishers, Inc.

Chapter 11

COMPARISON OF THIRTEEN METHODS FOR aDNA (1731-1841) EXTRACTIONS

Emese Juhász and *Károly Nagy*

Institute of Medical Microbiology, Semmelweis University, Budapest, 1089, Hungary

ABSTRACT

The aim of this Chapter is to compare thirteen different DNA extraction methods for the isolation of PCR amplifiable aDNAs from mummified soft tissues recovered from 1731-1841 (Vác, Hungary). Among the techniques and methods, the silicabased *guanidium thiocyanate* treatment and the phenol/chloroform extraction method combined with PTB (*N-phenacyl-thiazolium bromide*) were found to be powerful for aDNA isolation. These results might be indicative for Plant Archaeogenetists.

INTRODUCTION

In the past decades, several DNA extraction methods have been developed to resolve the difficulties deriving from the features of ancient samples. As DNA degradation begins immediately after the death of the organism by endogenous nucleases, DNA becomes severely damaged by oxidation, depurination (resulting in baseless sites), deamination (most often cytosine to uracil) and hydrolytic processes. Strand breaks and large numbers of interstrand and intermolecular crosslinks between reducing sugars and amino groups, due to alkylation and Maillard reactions, also occur in aDNA molecules [1, 2, 3, 4, 5]. Unidentified PCR inhibitors, a mix up of the nuclear and organelle aDNAs, and cross contaminations with current DNAs can also happen in aDNAs [1]. However, under fortunate conditions such as rapid desiccation, low temperature and high salt concentration, and also with circumstances when DNA becomes adsorbed into a mineral matrix, the speed of enzymatic and microbial

---

* E-mail: nagykar@net.sote.hu

degradation becomes slower and the endogenous nucleases can themselves become destroyed and inactivated. That is the reason why, in the case of animal tissues, the quality and quantity of aDNA isolated from teeth and bones is better than those from soft tissue [6].

MATERIALS AND METHODS

Mummified dried soft tissue samples were obtained from the mummies recovered in Vác from the period of 1731-1841 (Natural History Museum, Budapst, Hungary). In order to avoid cross DNA contaminations, the samples were handled with extreme precaution. Biopsies were carried out with sterile tools. Separated working areas, and laminar air flow hoods were cleaned thoroughly and irradiated with UV-light for one hour.

The thirteen DNA extraction methods were the following:

1. DNA extraction buffer contained 50mM NaH_2PO_4, 50 mM NaCl, 250mM Tris (pH 7.7), 10mM PTB (*N*-phenacyl thiazolium bromide), 50mM DTT, 5% Triton X-100 and 0.5mg/mL proteinase K.

    1a. 170-250mg tissues were digested in this buffer for 48 hours at 55°C. DNA was extracted three times by adding phenol/chloroform/isoamyl alcohol (25:24:1) followed by chloroform/isoamyl alcohol (24:1) extraction. aDNA was precipitated by 0.5 vol. 7.5M ammonium-acetate and 2 vol. absolute ethanol for 24 hours at -20°C. After centrifugation, the pellet was washed with 70% ethanol and allowed to air-dry. aDNA was dissolved in sterile distilled water.

    1b. aDNAs gained with the (1a) method were further purified by filtering the DNA through a Gentra column (Gentra Systems, Inc) in order to eliminate possible PCR inhibitors from the DNA solution.

    1c. 25mg tissues were digested in the extraction buffer for 24 hours at 55° C. Equal vol. phenol was added to the samples. After centrifugation, the supernatant was separated and equal vol. phenol/chloroform/isoamyl alcohol (1:24:1) was added, followed by chloroform/isoamyl alcohol (24:1) treatment. DNA was precipitated by 0.5 vol. 7.5M ammonium-acetate and 2 vol. isopropanol for 24 hours at -20° C. After centrifugation, the aDNA pellet was washed with 70% ethanol and allowed to air-dry. DNA was dissolved in sterile distilled water.

    1d. The phenol/chloroform/isoamyl alcohol (1:24:1) treatment was followed by precipitation with absolute ethanol instead of isopropanol.

2. The aDNA extraction was carried out by the silicabased method. Samples were digested in demineralisation buffer (0.5M EDTA pH8 and 1mg proteinase K/mL) at 56° C for 48 hours. Lysis buffer (10M guanidium thiocyanate (GuSCN; 0.1M Tris-HCl buffer pH6.4; 0.2M EDTA solution pH8.0; and Triton X-100) was added to the digested samples followed by incubation at 56° C for 2 hours. After centrifugation, 25μL silica solution was added to the separated supernatant and mixed for 1 hour. After centrifugation, the silica pellets were washed five times with washing buffer (10M GuSCN solution in Tris-HCl buffer pH6.4). Then, the silica was washed twice with 70% ethanol (-20° C) and once with acetone (-20° C). Silica samples were dried in a heat block (56° C) for 2 hours. aDNAs were eluted from silica by adding sterile

distilled water, and placed in a water bath (65° C) for 1 hour. Supernatants were expected to contain DNAs [7, 8].

3. Tissue samples were submerged in 0.5M EDTA for 48 hours at room temperature. Then 2 mg of proteinase K and 0.5 mL of 0.1M PTB was added and samples were incubated at 65° C for 14 hours. aDNA was extracted from the digested samples with phenol/chloroform/isoamyl alcohol (1:24:1).

    3a. Equal volume of phenol was added to the samples. After centrifugation, the supernatant was separated and equal vol. phenol/chloroform/isoamyl alcohol (1:24:1) was added, followed by chloroform/isoamyl alcohol (24:1). DNA was precipitated by 0.5 vol. 7.5M ammonium-acetate and 2 vol. isopropanol for 24 hours at -20° C. After centrifugation, the pellet was washed with 70% ethanol and allowed to air-dry. DNA was dissolved in sterile distilled water.

    3b. From the phenol/chloroform/isoamyl alcohol (25:24:1) extracted samples, aDNA was precipitated by absolute ethanol instead of isopropanol.

4a. Samples were treated with demineralisation and lysis buffer as in method #2. After centrifugation, the supernatant was filtered through a Genta column according to the manufactures' protocol (Gentra Systems, Inc).

    4b. The aDNA solutions gained with method #4a were further treated. 60μL aDNA solutions were incubated for 2.5 hours at 37° C in the dark with 440μL digestion buffer used in the method #1 without proteinase K. After this pre-treatment, aDNA was isolated with phenol/chloroform/isoamyl alcohol, followed by precipitation with ammonium acetate (0.5 volume of 7.5M) and ethanol (2 vol) for 3 days at -20°C.

    4c. The aDNA solutions gained with method #4a were further treated. 40μL aDNA solutions were incubated for 2.5 hours at 37°C in the dark with 400μL digestion buffer used in method #1 without proteinase K. Then, the solution was filtered through a Genta column.

5. aDNAs were purified with BioRobot EZ1 (Qiagen) automatic DNA isolating machine. Samples were digested in 350μL lysis buffer with 20μL 1M DTT, 10μL 440mM PTB and 400μg proteinase K. After overnight digestion, the samples were centrifuged and supernatants were collected. 1μL carrier RNA (Qiagen) was added to 200μL supernatant and the DNA isolation process was carried out by the machine.

6. Following the conventional phenol/chloroform/isoamyl alcohol, samples were digested in a buffer composed of Tris-EDTA buffer containing 100mM NaCl, 10mM Tris-HCl pH8, 25μM EDTA, 0.5% SDS, 0.1mg/mL proteinase K.

7. aDNA was extracted with dextran blue. Samples were digested in extraction buffer containing 0.5% N-laurylsarcosine-Na salt, 0.1 M EDTA and 100 mg/ml proteinase K and incubated overnight at 37° C, followed by centrifugation. 3.5 μL dextran blue (1 μg/μL), 250 μL 4M ammonium acetate and 500 μL 96% ethanol were added to 250 μL supernatant. DNA was precipitated at -70° C [9].

RESULTS AND DISCUSSIONS

aDNAs isolated by methods #1c and #2 were successfully amplified in PCR reactions in two samples of the four tested. Method #3a was successful in one of the two samples. Method #4a was unsuccessful resulting in a greenish smear (Maillard phenomenon) on the gel. aDNAs gained by methods #4b and #4c resulted in a definable sign in one sample of the five studied, and method #5 resulted in an indistinct band. Methods #1a, #1b, #1d, #6 and #7 resulted in no detectable aDNA bands on agarose gel after PCR amplification (Figure 1).

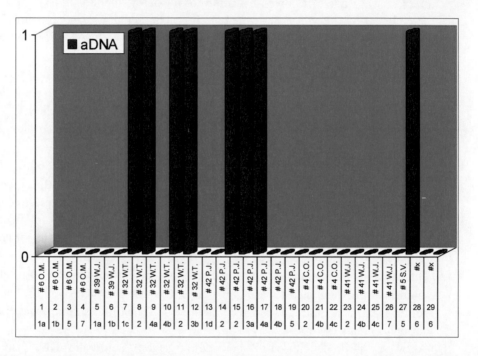

Figure 1. Comparative analysis of thirteen extraction methods (1a) to (7) of PCR-amplifiable aDNA samples (1 to 29) of eight individuals (#6 O.M. to #X) recovered from the 18th CENT. (1731-1841, Vác, Hungary).

CONCLUSIONS

Thirteen DNA extraction methods were tested on 29 tissue samples taken from eight mummies recovered from 1731-1841, in Vác, Hungary. Due to the chemical modifications and degradation of tissues, we faced significant difficulties in the extractions of aDNA. Since each archaeological site has peculiar characteristics and the level of aDNA damages may be different, the extraction method needs to be adjusted to the characteristics of the site where the remains were found [3]. In PCR amplifications of this study, the cycle numbers were increased from 35 to 40 cycles. With this change, nonspecific amplification products developed in the current DNA used for control, and also in some aDNAs, but without this increased cycle numbers, aDNA resulted in only faint bands. The PCR efficiency was also increased by the application of high fidelity polymerases, and the addition of uracil-N-glycosylase, which removes deaminated cytosine [1, 2].

In our study, amplifiable aDNAs were gained by five (#1c, #2, #3a, #4b; #5) of the thirteen methods tested. According to the criteria for aDNA study [2, 5, 10, 11], the extraction and preparation of PCR reactions were carried out in a rigorously separated laboratory to escape cross contamination with current DNAs. The silicabased guanidium thiocyanate treatment (method #2) proved to be the most powerful. However this method had a disadvantage of many washing steps that increase the chance of contamination even under extreme precautions. The common point in the four other methods (#1c, #3a, #4b and #5), which resulted in amplifiable aDNA, was the application of PTB (N-phenacyl thiazolium bromide), which appears to break intermolecular cross-links caused by glycosylation. The exact mode of action is unclear, nevertheless the effects of the Maillard reaction could be eliminated with the application of PTB [3, 4]. In methods #1c and #3a, the phenol/chloroform extraction was more effective when the initial phenol treatment was followed by the application of phenol/chloroform/isoamyl alcohol with the ratio 1:24:1, instead of a 25:24:1 ratio. Furthermore, isopropanol was better for aDNA precipitation than ethanol. Although methods #4b and #5 were successful in some cases, they were too complicated to use for serial aDNA extraction. For sampling, tissue sampled from deeper body sites (muscle, inner organ) were found to be better than those from superficial (skin) sites.

To conclude, silicabased methods combined with guanidium thiocyanate treatment (method #2) and methods #1c, #3a, #4b and #5 with the application of PTB (N-phenacyl-thiazolium bromide) were found to be powerful for aDNA isolation. These results might be indicative for Plant Archaeogenetists [12, 13, 14].

ACKNOWLEDGMENT

The authors say thanks to Ildikó Pap and Zsuzsa Guba (Department of Anthropology, Natural History Museum of Hungary, Budapest) for their contribution in the experiments, and Prof. L Waters (Auburn Univ, AL, USA) for his comments on the manuscript. The research was partially supported by the Hungarian Science Foundation (OTKA 61155) and the Doctoral School of Semmelweis University.

REFERENCES

[1]   Hofreiter M, Serre D, Poinar HN, Kuch M, Pääbo S. *Ancient DNA. Nat Rev Genet.* 2001 May; 2(5):353-9.

[2]   Pääbo S, Poinar H, Serre D, Jaenicke-Despres V, Hebler J, Rohland N, Kuch M, Krause J, Vigilant L, Hofreiter M. Genetic analyses from ancient DNA. *Annu Rev Genet.* 2004; 38:645-79.

[3]   Cipollaro M, Galderisi U, Di Bernardo G. Ancient DNA as a multidisciplinary experience. *J Cell Physiol.* 2005 Fcb; 202(2):315-22.

[4]   Willerslev E, Cooper A. *Ancient DNA. Proc Biol Sci.* 2005 Jan 7; 272(1558):3-16.

[5]   Hofreiter M, Jaenicke V, Serre D, Haeseler Av A, Pääbo S. DNA sequences from multiple amplifications reveal artifacts induced by cytosine deamination in ancient DNA. *Nucleic Acids Res.* 2001 Dec 1; 29(23):4793-9.

[6]   Baubliene J, Daugnora L, Miceikiene I. Evaluation of the DNA extraction method from ancient animal bones. *Ekologija* (Vilnius) 2003 Nr 1: 8-11.

[7]   Höss M, Pääbo S. DNA extraction from Pleistocene bones by a silica-based purification method. *Nucleic Acids Res.* 1993 Aug 11; 21(16):3913-4.

[8]   Boom R, Sol CJ, Salimans MM, Jansen CL, Wertheim-van Dillen PM, van der Noordaa J. Rapid and simple method for purification of nucleic acids. *J Clin Microbiol.* 1990 Mar; 28(3):495-503.

[9]   Kalmár T, Bachrati CZ, Marcsik A, Raskó I. A simple and efficient method for PCR amplifiable DNA extraction from ancient bones. *Nucleic Acids Res.* 2000 Jun 15; 28(12):E67.

[10]   Mitchell D, Willerslev E, Hansen A. Damage and repair of ancient DNA. *Mutat Res.* 2005 Apr 1; 571(1-2):265-76.

[11]   Marota I, Rollo F. Molecular paleontology. *Cell Mol Life Sci.* 2002 Jan; 59(1):97-111. Review.

[12]   Gyulai G, M Humphreys, R Lagler, Z Szabó, Z Tóth, A Bittsánszky, F Gyulai, L Heszky. Seed remains of common millet from the 4th (Mongolia) and 15th (Hungary) centuries: AFLP, SSR and mtDNA sequence recoveries. *Seed Science Research* 2006; 16: 179-191.

[13]   Lágler R, G Gyulai, M Humphreys, Z Szabó, L Horváth, A Bittsánszky, J Kiss, L Holly, L Heszky. Morphological and molecular analysis of common millet (*P. miliaceum*) cultivars compared to an aDNA sample from the 15th century (Hungary). *Euphytica* 2005; 146: 77-85.

[14]   Szabó Z, G Gyulai, M Humphreys, L Horváth, A Bittsánszky, R Lágler, L Heszky. Genetic variation of melon (*C. melo*) compared to an extinct landrace from the Middle Ages (Hungary) I. rDNA, SSR and SNP analysis of 47 cultivars. *Euphytica* 2005; 146: 87-94.

In: Plant Archaeogenetics
Editor: Gábor Gyulai, pp. 135-141

Chapter 12

COMPUTER-ASSISTED SEED MORPHOMETRY - A TOOL FOR ARCHAEOGENETICS

Irwin Rovner[*]

1902 Alexander Road, Raleigh, NC 27608, USA

ABSTRACT

The computer-assisted morphometry of two-dimensional and three-dimensional size, shape, and other morphological parameters are used in the analysis of seed morphology and variation to discriminate wild and domesticated population. An index RGiD (the *R*ovner-*G*yulai *I*ndex of seed *D*omestication) was developed (Rovner and Gyulai 2007) based on the comparative analysis of wild and domesticated forms of wheats discussed in this Chapter.

INTRODUCTION

Morphometry is a quantitative measurement of size and shape, which offers robust and operator independent parameters employed in statistical, analytical, and classification routines. Object morphometry and object recognition are achieved using five parameter categories: dimension of size, dimensionless ratio of shape, harmonic analysis, fractal dimension, and topology (Rovner 1971, 1994, 1995; Rovner and Russ 1989). The shape of a distribution plot derived from such measurements easily shows if the distribution is *continuous*, *normal* (i.e., Gaussian, bell-shaped), *non-parametric*, or *multimodal*. Morphological measurements are based on pixel counts occurring within the margin of each binary object multiplied by calibrated pixel size with adjustments for specific measurements (Figures 1, 2).

---

[*] E-mail: irovner@earthlink.net

All histograms in our study were standardized to an arbitrary 20-bin distribution using the minimum and maximum values of the range of variation as the distribution limits. Simple observation readily distinguished the diffused, multimodal distribution of the *wild taxa* from the near normal to hypermodal configuration of the *domestic taxa* in most of the histograms. Standard descriptive statistical summaries were assessed to determine which, if any, quantitative statistical measures could characterize and distinguish the distribution patterns. Standard statistical values (e.g., mean, variance, and standard deviation) were inconsistent and insensitive to the visually observable differences in distributions of size and failed to distinguish the wild and domestic populations. *Domestic* populations produced larger mean area values more often than those in *wild* populations supporting the conventional use of size differences to distinguish wild and domestic populations.

Figure 1. Samples of seed preparation for morphometry. LM photographs and digital images of *T. turgidum* ssp. *turanicum* (#8) seeds (a), digitalized images seeds (b), and together for size control (c) (scale, mm).

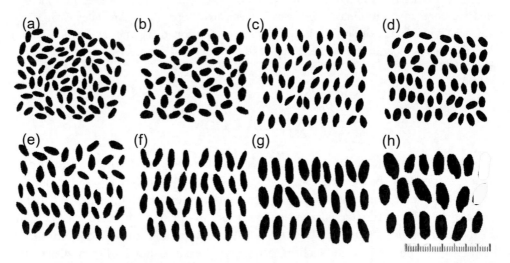

Figure 2. Digital images of *Triticum* seeds at the same magnification level (scale, mm) prepared for computer assisted seed morphometry. (a) *T. turgidum* (#9). (b) *T. aestivum cv. icterium* (#1). (c) T. *monococcum* ssp. *monococcum* (#4). (d) *T. aestivum* cv. *lutescens* (#2). (e) *T. aestivum* cv. *Transylvanicum* (#3). (f) *T. turgidum* cv. *turanicum* cv. *Kamut* (#5). (g) *T. turgidum* ssp. *turanicum* (#8). (h) T. *turgidum* ssp. *carthlicum* (#6). Samples in increasing size order shows an increment in endosperm hypertrophy.

We selected seed measurement data from the database of five taxa of domestic wheat: (1) *Triticum aestivum* cv. *icterium*, (2) *T. aestivum* cv. *lutescens*, (3) *T. aestivum* cv. *Transylvanicum*, (4) T. *monococcum* ssp. *monococcum*, and (5) *T. turgidum* cv. *turanicum* cv. *Kamut*. We also chose four wild wheat taxa: (6) T. *turgidum* ssp. *carthlicum*, (7) *T. dicoccoides,* (8) T. *turgidum* ssp. *turanicum*, and (9) *T. turgidum.*

RESULTS AND DISCUSSION

In morphometry, the situation is no different if the objects appear to be round in binary image. Euclidean measurement is derived from the snug round box the object fits into, as compared to size calculation based on calibrated pixel counts. In reality, it is rare that such objects are perfectly round, an aspect that can be determined by computer measurements of shape, such as Roundness. In general, Euclidean measurements of irregular objects are not direct measurements of the objects; instead, the measurements are of the nearest simple Euclidean polygon that fits around the object.

Shape factors are based on ratios of selected measurements of size and designed to be dimensionless, that is, size independent. The shape of every square, for example, is the same whether it is a large square or a small square. Perhaps the simplest shape factor to understand is Aspect Ratio, i.e., length divided by breadth, which measures an aspect of elongation of an object. Roundness measures the degree of departure from a circle of an object's two-dimensional binary configuration. This is based not on a visual image or an estimate of shape; rather, it is based on the mathematical fact that, in a circular object with a fixed area, an increase in the length of the object causes the shape to depart from a circle.

The *Roundness* calculation is constructed so that the value of a circle equals 1.0, while departures from a circle result in values less than 1.0 in direct proportion to the degree of deformation. Such quantitative objectivity can be enlightening as, for example, counter to intuitive perception, a square mathematically has a relatively high degree of roundness.

Formfactor is similar to Roundness, but emphasizes the configuration of the perimeter rather than the length relative to the object area. It is based on the mathematical fact that a circle (Formfactor value constructed to equal 1.0), compared to all other two-dimensional shapes (regular or irregular), has the smallest perimeter relative to its area. Since every object has a perimeter length and an area, this mathematical relationship can be used to quantify the degree to which an object's perimeter departs from that of a smooth circle, resulting in a value less than 1.0.

Curl, which is calculated as length divided by center line distance (a.k.a., fiber or skeleton length), measures the degree of departure of an object from a straight line, which usually is applied to irregular lines or long, narrow (squiggly) objects.

Solidity and Convexity are based on ratios of the area or perimeter of the true object to the area or perimeter, respectively, of a snug polygonal box fitted around the object. The degree of difference between the object and its fitted box is a quantitative measure of the degree of irregularity of the object; irregularities itself become a quantifiable aspect of morphology. Computer morphometry includes many more measurements of quantitative morphology, including surface texture, fractals, number of loops and branches, number of holes, hole size and shape, etc. (Russ 1990; Russ and Dehoff 2002; Willcox 2004).

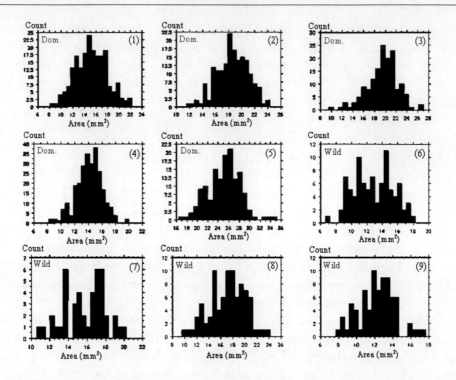

Figure 3. Area size distributions for five domestic and four wild wheat seed taxa. Note the near-normal distribution patterns for domesticates compared to multimodal distributions for wild taxa. Domestic Wheat Populations: (1) *Triticum aestivum* cv. *icterium*. (2) *Triticum aestivum* convv. *lustescens*. (3) *Triticum aestivum* cv. *Transilvanicum*. (4) *Triticum monococcum* ssp. *monococcum*. (5) *Triticum turgidum* cv. *turanicum* cv. *Kamut*. Wild wheat populations: (6) *Triticum turgidum* ssp. *carthlicum*. (7) *Triticum diccocoides* and *T. dicoccon* (8) *Triticum turgidum* ssp. *turanicum*. (9) *Triticum turgidum* (Rovner and Gyulai 2007).

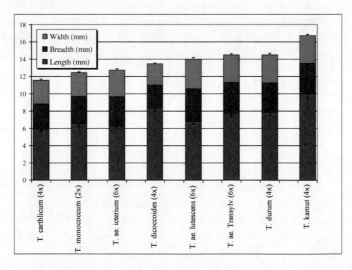

Figure 4. Due to the domestication events, the three cumulative seed parameters of wheat seed morphometry shows not a linear evolution in the endosperm hypotrophy from the diploid wheat (2n = 2x = 14) *T. monococcum*, through tetraploids (2n = 4x = 28) *T. carthlicum, T. dicoccoides, T. durum, T. Kamut* to the hexaploids (2n = 6x = 42) *T. aestivum* varieties.

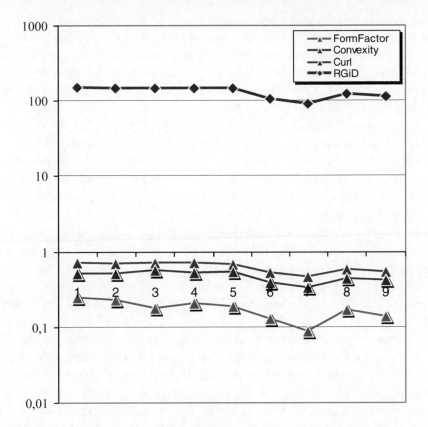

Figure 5. Seed morphometry and RGiD indexes of domestic (1-5) and wild (6-9) wheats. (1) *Triticum aestivum* cv. *icterium* (n = 203), (2) *T. aestivum* cv. *lutescens* (n = 169), (3) *T. aestivum* cv. *Transylvanicum* (n = 154), (4) *T. monococcum* ssp. *monococcum* (n = 264), (5) *T. turgidum* cv. *turanicum* cv. *Kamut* (n = 141); and (6) T. *turgidum* ssp. *carthlicum* (n = 44), (7) *T. dicoccoides* (n = 88), (8) *T. turgidum* ssp. *turanicum* (n = 87), and (9) *T. turgidum* (n = 79).

Calculation of RGiD index. Three Shape factors Formfactor, Convexity, and Curl provided comparative measurements that distinguished the shape patterns of wild versus domestic histograms (Figure 3). To simplify the results, the decimal points in each of the shape factor values were dropped and the three values added together to create a single index value, arbitrarily named RGiD - the Rovner-Gyulai Index of (seed) Domestication (Figures 4, 5).

CONCLUSIONS

Computer-assisted morphometric analysis of seeds appears to have exposed some fundamental problems with assumptions and conventional protocols applied to the study of morphological variation in seed taxa.

Nature is not normal. Size distributions of *wild* seed taxa do not typically conform to Gauss-Jan bell-shaped curves. Seed size distributions are usually *multimodal* rather than normal, and may reflect Darwinian adaptive variation factors where multiple modes represent competing and functional breeding populations within the larger species gene pool. This suggests that normal distributions, which are single mode, are likely to be highly specialized and narrowly varied in terms of natural selection. Single mode populations are therefore expected to be less flexible and less competitive in responding to adverse selection factors.

All five domestic wheat (*Triticum*) taxa had relatively high, tightly clustered RGiD values between 142 and 147, clearly separated from four wild wheat varieties with values from 92 to 120.

Specifically, three of the five domestic wheat seed populations (T. *aestivum lutescens*, T. *aestivum Transilvanicum*, and *Kamut*) exceeded the values for all wild populations, while two of four wild populations (subsp. *carthlicum* and subsp. *turgidum*) fell below all domestic populations. However, this size distinction is not consistent throughout the genus. The mean values of two of the four wild populations fell into the domestic range of values while two of the five domestic populations fell into the wild range of values (Figure 3).

The shift from a multimodal distribution of seed size (i.e., area) variation toward a more normal configuration is more likely to reflect genetic change because this shift appears to occur independent of changes in mean size values of replicate seed populations.

Domestic seed populations tend toward normal or hypernormal distributions indicating a higher degree of morphological specialization, likely due to human selection factors (Gyulai *et al.* 2006, 2009; Dane and Liu 2007).

A standard of 10 seeds as a representative sample is ineffective and unreliable. Mean values, central tendencies, arbitrary idealized type constructs are not supported as accurate or effective constructs to represent a taxon in order to study the nature of seed morphology. Mean values of size are inconsistent and highly variable. A comparatively larger mean size *sui generis* is not necessarily evidence of domestication; rather, it can also result from the inherent nature of variability in seed populations. Shape parameter mean values are also variable and inconsistent in seed populations. It appears that comparative analyses must be conducted at the level of population range of variation.

The configuration of population variation, e.g., the RGiD index, may provide a robust, independent means of analysis of seed morphology. It describes aspects of seed population variation important in comparative studies that are not expressed in conventional statistical description. This does not render conventional statistics meaningless; rather, each system seemingly describes something different in processing data from a given seed population. Thus, seed analysis should be enhanced through a greater understanding of what the two systems are expressing and then applying both in a complementary fashion. For example, a seed population that exhibits both increased size and a high RGiD value increases the confidence of an assessment of a domestic population.

It is also interesting to note the difference between tight clustering of domestics and the more varied results evident in the wild populations. This may again directly reflect the nature of variation. Perhaps most significant, is the fact that there is no correlation between the RGiD values and mean size values for any of the wheat populations, wild or domestic. They are independent reflecting different aspects of population morphology and variation.

The RGiD Index is a new approach to morphogenetical (phenotypic) analysis utilizing computer-assisted morphometry. Like all new approaches, it must be tested further and its accuracy verified. One of the apparent benefits is the fact that this index virtually abandons the typological concept in favor of an assessment of objectively measured morphological variation.

Assuming the transition to domesticated forms occurred over a substantial period of time, archaeological seed assemblages, carefully collected and dated, could be used to trace the transition from multimodal to hypermodal configuration, tracking the process of domestication itself. Thus, RGiD values might be used to determine the rate and duration of the domestication process through time on a taxon-by-taxon basis.

REFERENCES

Dane F, J Liu (2007) Diversity and origin of cultivated and citron type watermelon (*Citrullus lanatus*). *Genet Resour Crop Evol* 54:1255-1265.

Gyulai G, M Humphreys, R Lágler, Z Szabó, Z Tóth, A Bittsánszky, F Gyulai, L Heszky (2006) Seed remains of common millet from the 4th (Mongolia) and 15th (Hungary) centuries: AFLP, SSR and mtDNA sequence recoveries. *Seed Science Research* 16: 179-191.

Gyulai G, Z Tóth, Z Szabó, F Gyulai, R Lágler, L Kocsis, L Heszky (2009) Domestication Events of Grape (*Vitis vinifera)* from Antiquity and the Middle Ages in Hungary from growers' viewpoint. *Hung Agric Res* 2009/3-4: 8-12.

Rovner I (1971) Potential of Opal Phytoliths for Use in Paleoecological Reconstruction, *Quaternary Research* 1: 345-359.

Rovner I (1994) Floral history by the back door: Phytolith analysis of two residential yards at Harpers Ferry." *Historical Archaeology* 28: 37-48.

Rovner I, F Gyulai (2007) Computer-assisted morphometry: A new method for assessing and distinguishing morphological variation in wild and domestic seed populations. *Economic Botany* 61: 154-172.

Russ JC (1990) Computer-assisted microscopy: The measurement and analysis of images Plenum Press. New York.

Russ JC, RT DeHoff (2002) Practical stereology 2nd editionPlenum Press. New York.

Russ JC, I Rovner (1989) Stereological identification of opal phytolith populations from wild and cultivated *Zea*. *American Antiquity* 53:784-792.

Willcox G (2004) Measuring grain size and identifying Near Eastern cereal domestication: Evidence from the Euphrates valley. *Journal of Archaeological Science* 31:145–150.

In: Plant Archaeogenetics
Editor: Gábor Gyulai, pp. 143-159

ISBN: 978-1-61122-644-7
© 2011 Nova Science Publishers, Inc.

Chapter 13

PLANT ICONOGRAPHY - A SOURCE OF INFORMATION FOR ARCHAEOGENETICS

Jules Janick[1],*, Marie-Christine Daunay[2]
and Harry S. Paris[3]

[1]Department of Horticulture and Landscape Architecture, Purdue University,
West Lafayette, Indiana 47907, USA
[2]INRA, UR 1052 Unité de Génétique et d'Amélioration des Fruits et Légumes,
Montfavet 84140, France
[3]Department of Vegetable Crops and Plant Genetics, Agricultural Research Org., Newe
Ya'ar Research Center, Ramat Yishay 30-095, Israel

ABSTRACT

Artistic works from prehistory and antiquity to the present constitute an alternate source of information on crop plants that become a valuable resource for investigations involving genetic and taxonomic information, as well as crop history. Sources of plant iconography include cave paintings, ancient mosaics, sculpture, carvings and inlays, frescos, tapestries, illustrated manuscripts, herbals, and books, as well as photographs of desiccated remnants.

INTRODUCTION

Works of art from pre-history and antiquity to the present, constitute an alternate source of information on plants and crops (Janick, 2007). Plant iconography becomes a valuable resource for investigations involving genetic and taxonomic information, as well as crop history, including evolution under domestication, crop dispersal, lost and new traits. Crop images are one of the unequivocal tools for assessing the historical presence of botanical taxa

* E-mail: janick@purdue.edu

in a particular region and are an especially valuable resource for determining morphological changes of crops from antiquity to the present (Gyulai *et al.* 2006, 2009; Dane and Liu 2007). Often early written descriptions are ambiguous and the confusion of plant names in ancient documents makes the image the "smoking gun" in determining the precise species involved, plus providing information on the presence of morphological characters that may be unclear from the text. Although a plethora of ancient plant images exists, they are widely scattered among libraries and museums, and are often difficult to locate and to access. Moreover, images are often copyrighted and costs for access tend to be high. Many museums and libraries restrict photography and restrict viewing of rare books, making research a difficult, time consuming and expensive activity. However, the digitization of information by some of the major world libraries has greatly facilitated the search for ancient illustrations, although they still remain expensive to publish. An attempt has been made to assemble illustrations of cucurbits and solanaceous crops (Janick *et al.*, 2006, 2007b). The purpose of this chapter is to review the role of crop iconography in understanding crop history and to introduce some sources for archaeogenetics.

There are numerous sources of plant iconography. These include cave paintings, ancient mosaics, sculpture, carvings and inlays, frescos, tapestries, illustrated manuscripts, herbals, and books, as well as photographs of desiccated remnants. In this paper, we will briefly review sources of crop images from a historical perspective. Examples of studies on crop iconography by the authors and colleagues are included in the References section (Daunay and Janick, 2007, 2008; Daunay *et al.*, 2007ab, 2008, 2009; Hummer and Janick, 2007; Janick 2002, 2004a,b, 2007; Janick and Caneva, 2005; Janick and Daunay, 2007; Janick and Paris, 2005, 2006a,b,c; Janick *et al.*, 2006, 2007a,b, 2009, 2010; Paris and Janick, 2005, 2008a,b; Paris *et al.*, 2006a,b, 2009; Renner *et al.*, 2008).

EPOCHS

The Paleolithic and Neolithic Record

The Paleolithic Era (Old Stone Age) extends back 2.5 million years when primitive humans were hunters and gatherers and used stone tools, to the introduction of agriculture about 12,000 BP known as the Neolithic Era. Their culture survived chiefly though the remains of stone tools, whose features along with carbon dating can be used to determine chronology.

About 500,000 years ago, a surge of hominid expansion occurred from populations that had a brain capacity of 1100–1300 cc, a species now known as archaic *Homo sapiens*. This group gave rise to Neanderthals; brawny, large brained, tool-making humans that appeared in Europe about 250,000 years ago but were destined to be overtaken by an African group via Asia, *Homo sapiens* or Cro-Magnon, the human species that is us. About 200,000 years ago, these two groups coexisted but Neanderthals disappeared about 25,000 to 30,000 years ago. Legacies of this new group, in the form of cave paintings, emphasized the hunt and represent the beginning of art history. The images were mostly of animals but included plants (Tyldesley and Bahn, 1983) (Figure 1). Some sculptures of voluptuous women (Figure 2) known as Venuses, still present an emotional impact related to the keen interest of early humans in fertility, but evidence of clothing made from plant and animal sources indicates the development of weaving and textile technology. The spread of agriculture about 10,000 to 12000 BP occurred in a relatively short period of time and this "sudden" transformation in culture is referred to as the Neolithic Revolution. Crop plants can be identified in Neolithic and Bronze Age images (Figure 3, 4).

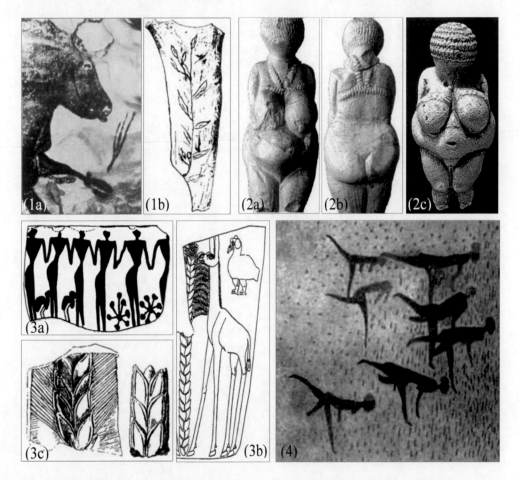

Figures 1- 4. The Paleolithic and Neolithic Records. Figure 1. Paleolithic images of plants (17 - 30,000 years BCE): *(a)* aurock with a primitive image of a plant; *(b)* a more sophisticated image showing stem and leaves carved on a reindeer horn (Source: Tyldesleay and Bahn, 1983). *Figure 2*. Woman figurine of the Paleolithic period *(a, b)* showing evidence of textile technology. (source: Janick, 2007); *(c)* Woman figurine ('Venus' from Villendorf) discovered in 1908 by Hungarian archaeologist József Szombathy at a paleolithic site (ca. 25,000 years BCE) near Willendorf, Austro-Hungary. *Figure 3*. Neolithic and Bronze age representatives of crops where plants can be identified: *(a)* pottery image form Tejpe Sialk, Iran, 7000 BCE; *(b)* predynastic Egyptian image of palm tree and gazelle 5000 BP; *(c)* cereal carvings, Egyptian first dynasty 5000 BCE (Source: Janick, 2007). *Figure 4*. Women gathering grain, Tassilil n'Ajjer, Algeria (5000-6000 BCE) (Source: Musée de l'Homme, Paris).

Mesopotamia

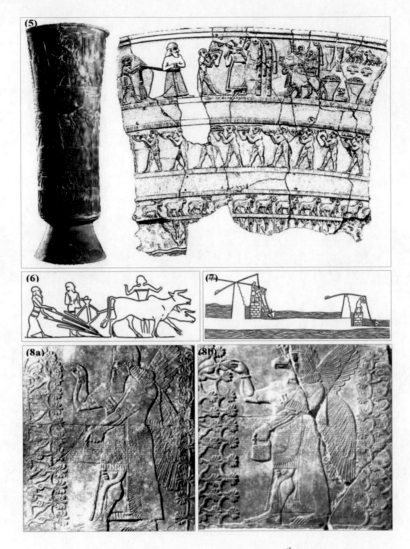

Figures 5-8. Mesopotamia. Figure 5. The Uryuk vase (ca, late 4th millennium BCE, Attendants offer fruit in a wedding ceremony, probably between a priest king and the goddess Innana (Istar). Note the presence of barley and sesame at the bottom next to the river (Source: Pollack, 1999). *Figure 6.* A Babylonian scratch plow with seed drill from a Cassie cylinder seal, 2nd millennium BCE (Source: Singer *et al.*, 1954). *Figure 7.* Raising water from the river with a shaduf by Assyrians. From the palace of Sennacherib at Nineveh, Mesopotamia, 7th century BCE (Source: Singer *et al.*, 1954). *Figure 8a,b.* Date palm pollination depicted in Assyrian bas reliefs, 883–859 BCE. The pollinator assumes the form of a godlike figure (genie) and the date palm has been transformed into a symbolic tree (Source: Paley, 1976).

The ancient Near East cultures known as Mesopotamian civilization are largely based on Semitic populations that existed between the Tigris and Euphrates Rivers that soon expanded to the area known as the Fertile Crescent, which includes parts of present day Israel, Jordan, Lebanon, Syria, Iraq, and Iran. A second Neolithic Revolution between 6000 and 3000 BCE (the Bronze Age), involved the change from villages to urban centers and the development of a settled agriculture coinciding with the beginning of fruit culture. This is well documented in the decorations of a vase, (Figure 5), late 4th millennium BCE, found in Uryuk (biblical Erekh), an ancient city on the Euphrates north of present-day Basra, Iraq, that is associated with Sumerian civilizations, where writing was invented. Other evidence of agricultural technology includes the refinement of a plow with a seed drill (Figure 6), date palm pollination (Figure 7), and irrigation technology (Figure 8).

Ancient Egypt

Paleolithic-Neolithic artifacts along the Nile date back 12,000 years. Nubian, Ethiopian, and Libyan populations fusing with Semitic and West Asian immigrants formed a people and created the Egypt of history. A continuous 6000 year record exists with a unique and productive agriculture at its base. Knowledge of the history of Egyptian agronomy and horticulture can be gleaned from the archeological record and is supported by surviving written Egyptian documents, temple inscriptions, tombs paintings, as well as commentary from Antiquity such as the Hebrew Bible, the Greek historian, Herodotus (484–413 BCE), and the Greek philosopher and botanist, Theophrastus (372–288 BCE).

The ancient technology of agriculture can be vividly reconstructed from the artistic record, paintings and sculpture in tombs and temples dating onward from 3000 BCE. Agricultural activities were favorite themes of artists, who illustrated lively scenes of daily life that adorn the tombs of the pharaohs and dignitaries. The artistic genius engendered by Egyptian civilization, the superb condition of many burial chambers, and the dry climate have made it possible to reconstruct a detailed history of agricultural technology. Ancient Egypt is shown to be the source of much of the agricultural technology of the Western World. Illustrations of these artifacts can be gleaned in four key references: Keimer (1924), Singer *et al.* (1954), Darby *et al.* (1977), and Manniche (1989). The Museum of Agriculture in Cairo has a rich collection of artifacts, including desiccated plant specimens.

Examples of the presence of plant images from ancient Egypt are shown in a brief sampling of the artistic record. This includes harvest of pomegranates, grape harvest and wine making (Figure 9), and a collection of cucurbits (Figure 10). The absence of images of cucumber (*Cucumis sativus* L.) supports the conclusion that the many reference to cucumbers in English translations of ancient texts should be understood as snake melons, *Cucumis melo* L. subsp. *melo* Flexuosus group (Janick *et al.*, 2007a).

Pre-Columbian America

Aztec, Maya and Inca cultures in the Americas at the time of the encounter of Columbus with the New World, were monumental civilizations similar in many ways to that of ancient Egypt of 2000 BCE, with a developed agriculture, enormous temples in the form of pyramids, pictorial writing, a system of cities and government, a bewildering theology, and magnificent art.

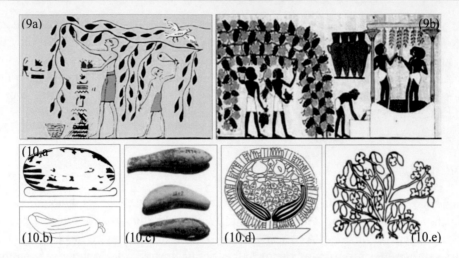

Figures 9-10. Ancient Egypt. Figure 9. Fruit harvest in ancient Egypt: *(a)* pomegranate harvest and a boy using a sling to chase birds; *(b)* wine production with collection of grapes under a pergola (Source: Janick, 2007). *Figure 10.* Cucurbit images in ancient Egypt: *(a)* watermelon (*Citrullus lanatus*) showing oblong striped fruit, Old Kingdom, *(b)* snake melons (*Cucumis melo* subsp. *melo* Flexuosus Group) with striations, with attached peduncle and corolla, 18th Dynasty; *(c)* a basket of striped melons and dates from a wall painting at Thebes, 18th Dynasty, *(d)* wooden models of melon, New Kingdom; *(e)* carving of squirting cucumber (*Ecballium elaterium*) based on thick root, bushy growth habit, and small oval in an image of an early botanical collection on the walls of the temple of Karnak ca. 1450 BCE (Source: Janick *et al.,* 2007a).

There was a dark side, too, of human sacrifice, cannibalism, slavery, and constant warfare. The rich art of these civilizations include embroideries (Figure 11) and ceramics that celebrate the domestication of indigenous crops (Figure 12). Unfortunately, many of the codices were destroyed by the conquistadores and the church. Native artists post-Columbus made detailed illustrations of potato culture in the form of a calendar (Figure 13) as gifts to the King of Spain.

SOURCES

Mosaics and Inlays

The assemblage of images from small pieces of colored glass, stone, or gems referred to as mosaics, date to the third millennium BCE. Mosaics were popular in ancient Greece and Rome and survive in Christian and Islamic art up to the present. Mosaics were prominent as decorations on floors, walls, and ceilings of private residences and public buildings, especially churches, mosques, palaces or mansions and constitute some of the glories of ancient, medieval, and Renaissance art in the West. Many thousands of ancient mosaics have been preserved and are actively studied by art historians and can be accessed online. Mosaic art spread throughout the Roman Empire and are particularly rich in areas that today are in Italy, Tunisia, Libya, Syria, and Turkey. The Hatay (Antakya) Archeological Museum has a splendid collection from the ancient city of Antioch of ancient Syria (Hatay province, however, is now part of Turkey) based on discoveries in 1932–1939.

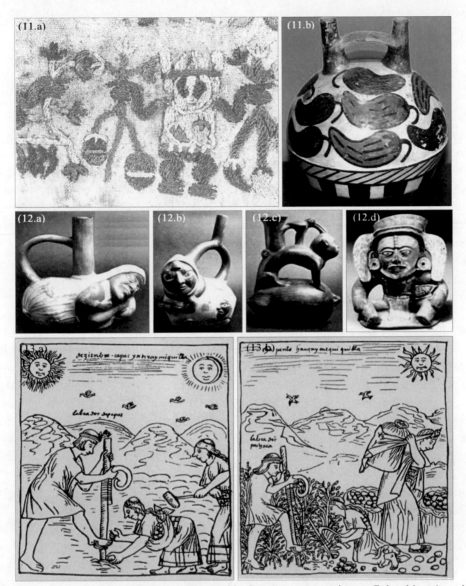

Figures 11-13. Pre-Columbian America. Figure 11. Capsicum pepper in pre-Columbian America: *(a)* embroidery showing fruits held by a man and two fruits in chords around his neck, 400-500 CE; *(b)* pottery from Peru, Mochica period (1–600 CE) (Source: Daunay *et al.,* 2008). *Figure 12.* Pre-Columbian ceramic jars from Peru: *(a)* peanut; *(b)* potato; *(c)* squash; *(d)* cacao pod (Source: Leonard, 1973). *Figure 13.* Planting and harvesting of potato by the Incas (1580) (Source: Leonard, 1973).

Roman mosaics include rich scenes of plants and agricultural crops. One panel (Figure 14) from the 3rd century CE depicts agricultural scenes and contains the first image of grafting along with other scenes of fruit culture. Examples of cucurbits and other fruits are shown in Figure 15.

Mughal mosaics are found among the decorations of the Taj Mahal, constructed in Agra, India, by Shah Jahan from 1632 to 1658 as a memorial to his wife known as Mumtaz Mahal (Janick *et al.* 2010). Islamic decoration restricts graven images of humans but is rich in botanical subjects and includes floral inlays known as *pietra dura* (Figure 16) and sculpted bas reliefs known as *dados* (Figure 17). The plant images are rich in ornamental geophytes (bulb crops) common to the region.

Sculptures

Plants in sculptured form are found in Egyptian, Greek, Roman, Indian and Renaissance art. In ancient Egypt, the papyrus and lotus were symbols of the upper and lower Nile region; and the reunification of Egypt in the third millenmium BCE is shown in illustrations where these two plants are intertwined (Figure 18.a) and these forms are also found in architectural columns (Figure 18B). A Roman bas relief of snake melon from Merida, Spain is identified by its leaves and striated fruit (Figure 19). The cathedral bronze doors in Pisa, Italy dated 1601, are rich in sculpted food crops that surround the panels of religious scenes and include eggplant, cucumber, and a large, ribbed, tomato (Figure 20).

Illustrated Manuscripts

The *Juliana Anicia Codex (JAC)* or *Codex Vindobonenis*, 512 CE, is a magnificent, illustrated manuscript from late antiquity found in Constantinople and based on the famous herbal *Peri Ylis Iatrikis* (Latinized as *De Materia Medica, On Medical Matters*) originally written about 65 CE by the Roman army physician Pedanius Dioscorides (20–70 CE) born in Anazarbus, Cilicia, in what is now southeastern Turkey. The manuscript, made for the daughter of the Roman emperor Anicius Olybrius, contains descriptions, medical uses, and illustrations of almost 400 plants listed alphabetically and can now be accessed through a two-volume facsimile edition, *Der Wiener Dioskurides* (1998, 1999). There is an English translation by Beck (2005) of the reconstructed, original, non-alphabetic Dioscoridean manuscript in German by Wellman (1906–1914). Four examples of crop information from the *JAC* are included in Figure 21.a,b,c, and 22.a. The painting of cowpea (*Vigna unguiculata*) shows indeterminate plant growth, an interesting genetic variant in this species. The illustration of two brassicas show the non-heading trait in cabbage and rather typical morphology of turnip. Finally, an elegant depiction of blackberry (Figure 21.a) shows primocane fruiting and rooting of shoots but there are botanical inaccuracies in leaf morphology and petal number (Hummer and Janick 2007). A similar drawing (Figure 22.b) in a later recension (ca. 675) of Dioscorides called *Codex Neapolitanus*, although not as fine as that of the *JAC*, is botanically more accurate in reference to leaf and petal morphology, indicating that both drawings derive from a lost template.

A late medieval example of crop illustrations can be found in a series of lavish versions of manuscripts known as the *Tacuinum Sanitatis* (*Tables of Health*), which were probably prepared as royal gifts in Europe. There are six major works (one is divided) in libraries in Liège, Vienna, Rome, Paris, and Rouen, which were commissioned by northern Italian nobility during the last decade of the 14th century and the course of the 15th century (Paris *et al.*, 2009; Daunay *et al.*, 2009; Janick *et al.*, 2009). The text is based on an 11th-century Arabic manuscript, *Taqwim al-Sihha bi al-Ashab al-Sitta* (*Rectifying Health by Six Causes*), written as a guide for healthy living by the Christian Arab physician known as Ibn Butlan (d.

Figures 14-17. Mosaics and Inlays. Figure 14. Roman pomology in mosaics from St. Roman-en-gal, 3rd century CE: *(a)* detached scion grafting; *(b)* fruit harvest; *(c)* juice extraction (Source: Janick, 2007). *Figure 15.* Cucurbits in Roman mosaics; *(a)* long fruited melon (*Cucumis melo* Flexuosus Group) with small yellow-orange corollas clinging to the acute stylar end from Tunisia, 2nd century; *(b)* immature and mature long fruited melons showing fruit splitting. Tunisia 3rd century; *(c)* round fruited melon *Cucumis melo* showing striping, Tunisa. 4th century; *(d)* bottle gourd (*Lagenaria siceraria*) showing characteristic swelling on the pedunular end; *(e)* youth holding bottle gourd in right hand and watermelon (*Citrullus lanatus*) in left hand and, in mosaic named August, Tegea Episkopi, late 4th to 5th century Peloponnese (Source: Janick *et al.,* 2007a). *Figure 16.* Floral inlays in the Taj Mahal (1632-1658): chrysanthemum (*Chrysanthemum*), central flower, plus Siroi lilies (*Lilium mackliniae*) on either side. (Source: Janick *et al.,* 2010). *Figure 17.* Floral bas relief in the Taj Mahal (1632–1658): (dado) of cut stem in an urn include iris in center and in descending order are columbine (*Aquilegia* sp.), daffodil (*Narcissus* sp.), columbine, windflower (*Anemone* sp.), tulip, (*Tulipa* sp.) windflower, capsule of poppy (*Papaver* sp.), delphinium (*Delphinium* sp.) (Source: Janick *et al.,* 2010).

Figures 18-20. Sculptures. Figure 18. Sculptured plants in ancient Egypt. *(a)* Intertwining of lotus and papyrus symbolizing the reunification of upper and lower Egypt. Source Cairo Museum; *(b)* The Ptolomeic Temple of Khnum (Kom Ombo) at Esna (2nd century BCE) showing columns representing papyrus and lotus (Photo by J. Janick). *Figure 19.* Marble relief entitled summer from Merida, Spain 4th century with sheaf of wheat and melon (*Cucumis melo* Flexuosus Group) showing dentated leaf and striated fruits (Source: Janick *et al.,* 2007a). *Figure 20.* Sculpted images of eggplant *(a)*, cucumber *(b)*, and tomato *(c)* in the bronze doors of the Cathedral in Pisa (1601) (Photo by J. Janick)

1063). Vivid agricultural imagery includes scenes of the harvest of vegetables, fruits, flowers, grains, and culinary and medicinal herbs, accompanied by a brief summary of the health aspects of the subject. Each of the manuscripts are drawn by different artists but are obviously related. The *Vienna codex Ser. N. 2644* contains the most accurate depictions, which include 9 cereals, 26 vegetables, 33 fruits, 3 flowers, and 21 culinary and medicinal herbs. The illustrations show crops at the optimal state of maturity and, moreover, are a rich source of information on life in the feudal society, with nobles engaged in play and romance while laborers work on the estate. A selection of crops from two versions of *Tacunum Sanitatis* is presented in Figure 23.

A French Royal prayer book, known as *Les Grandes Heures d'Anne de Bretagne* (*Manuscript Latin 9474*), contains prayers with illustrated margins, and full page monthly calendars and paintings of religious themes (Paris *et al.*, 2006). This stunningly illustrated manuscript was prepared for the personal use of Anne de Bretagne (1477–1514), twice Queen of France as wife of Charles VIII and Louis XII, by the famous artist Jean Bourdichon (1457–1521), probably painted between 1503 and 1508, about a decade from the return of Columbus to Spain. There are miniature paintings of plants and small animals, mostly insects, on each page that can be searched on www.hort.purdue.edu/newcrop/bilimoff/default.html. Well over 300 plant species are included. This work contains the first European illustration of a non-esculent gourd of *Cucurbita pepo* subsp. *texana* (Figure 24). The seed source for this gourd cannot be determined but could have been obtained from various sources. Seeds from the voyages of Columbus were transmitted in 1494 by Peter Martyr D'Angheria, Tutor to the Spanish royal household, to Cardinal Asconio Sforza secretary of state to the Vatican and seed could have reached France this way. Other possibilities include the voyages of Europeans, including Amerigo Vespucci who entered the Gulf of Mexico as early as 1498, or from various Bretons or Norman who reached the Americas by 1503, returning with parrots and Brasilwood.

Printed Herbals

Herbals, botanical works emphasizing medical uses of plants, are one of the most important sources of plant iconography (Eisendrath, 1961). A splendid introduction to the field can be found in Agnes Arber's 1938 book on herbals. While many herbals can be found in specialized library collections, they are scattered and difficult to locate and access, as a result of their value. Moreover, their nomenclature is not consistent and there is often a lack of adequate indices. Facsimile editions exist for a number of printed herbals, including the 1542 herbal of Leonhard Fuchs (Meyer *et al.*, 1999), the 1597 edition of the *Herball* of John Gerard(e) (Theatrum Orbis Terrarum) and the 1633 edition of Gerard amended by Johnson (Dover Publ.). The illustrations of Renaissance herbals are derived mostly from woodcuts and sometimes from original painted drawings. However, many herbals copy parts of text and woodcuts from previous herbals or are based on an exchange of woodblocks by printers, thus they often contain errors in identification. Woodcuts of sweet potato (*Ipomoea batatas*) and potato (*Solanum tuberosum*), the first printed illustration of potato in Europe, from the famous English *Herball* of John Gerard(e) (1597) are presented in Figure 25. Gerard is responsible for the confusion between potato, *Solanum tuberosum* (Indian name *papas*) and sweet potato, *Ipomea batatas* (Indian name *batatas*), because he labeled his printed

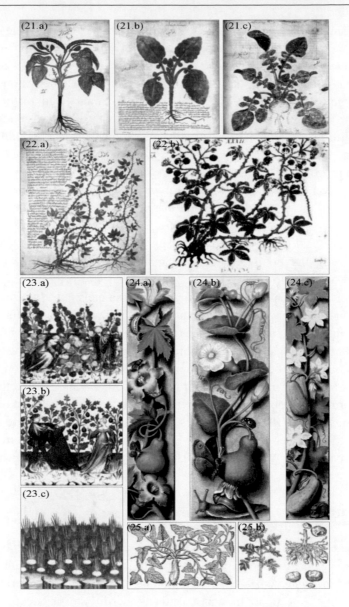

Figures 21-25. Illustrated Manuscripts. Figure 21. Crop illustrations from the *Juliana Anicia Codex* of 512 CE: *(a)* cowpea (*Vicia unguiculata*) showing indeterminate growth habit. *(b)* non-heading cabbage (Brassica oleracea) (c) turnip (Brassica rapa) (Source: Der Wiener Dioskurides, 1998, 1999). Figure 22. Blackberry (Ruus ulmifolius): (a) Juliana Anicia Codex; (b) Codex Neapolitanus. Figure 23. Crops of the Tacuinum Sanitatis, 14th century. (a) melon (Cucumis melo subsp. melo) from Vienna 2644, with golden large round fruit that are obviously aromatic as one is being sniffed by the courtier in the red gown, Vienna 2644; (b) aubergine / eggplant (Solanum melongena) with developing globose, purple fruit borne near plant apices is shown behind a fondling couple being admonished by a lady, implying that eggplant has aphrodisiacal properties, Vienna 2644 (c) Onion (Allium cepa) from showing red and white bulbs, Roma 4182. (Source: Paris et al., 2009; Daunay et al., 2009). Figure 24. Cucurbits of the Les Grandes Heures d'Anne de Bretagne (1503–1508): (a) Quegourdes de Turquie (Cucurbita pepo); (b) Quegourdes (Lagenaria siceraria); (c) Concombres (Cucumis sativus) (Source: Paris et al., 2006a). Figure 25. Woodcuts of (a) sweet potato (Ipomoea batatas) and (b) potato (Solanum tuberosum) from the 1597 Herball of John Gerard(e).

illustration of potato, the first one to be published in Europe, *Battata Virginiana sive Virginianorum & Pappus*, *Potatoes of Virginia*; Virginia being the area where the tubers he grew in his garden came from. A study of the iconography of the Solanaceae (Daunay *et al.*, 2008) shows the richness of information found in herbals. In the 17th and 18th centuries, botanical art became a sensation and many Royal collections of plant images called florilegias were made for their sheer beauty and for conveying the knowledge of exotic plants brought back by travelers around the world. They became the source of floral art for commercial uses such as fabrics and wallpaper. A useful list of florilegias and botanical codices can be found in Wikipedia.

Paintings

Paintings from antiquity to the present have often used plants and crops as themes for their aesthetic and/or symbolic value. Frescoes, paintings on flesh plaster, on walls and ceilings are well- preserved since the pigments seep into the plaster. The frescoes of Pompeii and Herculaneum in Italy have been preserved as a result of the eruption of Vesuvius in the year 79 and are valuable resources for ancient depictions of plants. Examples include images of figs and peach from Pompeii (Figure 26).

Paintings of plants increased during the Italian Renaissance. The Roman residence (now known as Villa Farnesina) of the wealthy Roman financier Agostino Chigi, decorated between 1515 and 1518, is a splendid source of crop images. The ceiling of the Loggia of Cupid and Psyche illustrate scenes from *Metamorphoses* (*The Golden Ass*) by Apuleius, a 2nd century CE Roman author, painted in fresco by Raphael Sanzio and his assistants, including Giovanni Martini da Udina, who was responsible for the festoons that are a fantastic source of crop images. The thousands of images of 163 species in 49 botanical families include some of the first illustrations of New World plants (Janick and Caneva, 2005; Janick and Paris, 2006a). The site can be navigated on http://www.hort.purdue.edu/newcrop/udine/default.html where individual species can be accessed. Included are the first images of maize, showing three distinct phenotypes (Figure 27 a,b,c). The different traits include ear shape, row number, silk color, and the presence of a phenotype called tassel seed in the short ears; this same trait can be shown in a pre-Columbian stone carving (Figure 27.d) and a photograph from primitive maize (Figure 27.e).

A genre of Baroque paintings known as still life (*natura morta*), emphasizing fruits vegetables, and flowers, is a rich source of information from the 17th and 18th centuries (Zeven and Brandenburg, 1986). A classic painting (Figure 28) by Michelangelo Merisi, also known as Cavaraggio, is considered the first fruit portrait, and contains evidence of damage inflicted by various diseases and pests (Janick. 2004a). Baroque painters found scenes of everyday life intriguing subjects to paint, and fruit and vegetable markets increasingly became a common subject. Two example of fruit market paintings are shown in Figure 29 a,b). The Flemish painter Pieter Aertsen painting entitled the *Produce Seller* (1567) is rich in *Brassica* crops including head cabbage (7 green and 1 red) as well as cauliflower and various cucurbits including bottle gourd, melon, pumpkin, and cucumber and also includes Belgium waffles! *The Fruit Seller* by Vincenzo Campi (1580) displays a plethora of fruits and vegetables in Italian markets included in the upper right a box of pears and young squash, *Cucurbita pepo* subsp. *pepo* Cocozelle group, with flowers attached, still a common commodity in Mediterranean countries (Janick and Paris, 2005; Paris and Janick, 2005; Tóth *et al.* 2007). Other noteworthy painters of crop images include Giovanna Garzoni (1600–1679) and Bartolomeo Bimbi (1648–1723).

Figures 26-29. Paintings. Figure 26. Roman frescoes from Pompeii (1st century CE): *(a)* basket of purple and green figs; *(b)* large, green, and freestone peach (Source: Jashemski, 1979). *Figure 27.* Images of maize in the Villa Farnesina (1515-1519): *(a)* long ears; *(b)* middle sized ears; *(c)* short ears. The tassel seed character in the short ears can be compared with *(d)* Aztec image and *(e)* photographs of primitive maize from Peru. Source: Janick and Caneva, 2005. *Figure 28.* Caravaggio's painting entitled *Still life of a Basket of Fruit* (1601) shows various disease symptoms including anthracnose on a Figure leaf, scab on a quince leaf, codling moth injury in a apple, Oriental fruit moth damage on a peach leaf, leaf roller damage on a pear, grape mummies, and grasshopper injury on grape leaves (Source: Janick, 2004a). *Figure 29.* Baroque market scenes: *(a)* Produce Seller (1567) by Pieter Aertsen (source; Honig, 1998). *(b)* The Fruit Seller (1580) by Vincenzo Campi (Source: Paris and Janick 2005).

CONCLUSIONS

This chapter of plant iconography should be considered a very brief introduction to the subject. The collection of images of individual crops combines the field of art, history, and crop evolution and genetic diversity. An analysis of the iconography of the *Solonaceae* (Daunay and Janick, 2007; Daunay *et al.*, 2008) suggests the various difficulties involved for the interpretation of the illustrated information. Plant iconography is an outstanding resource for research on crop history and diversity but unfortunately, the source of many of these works and in particular those from the East are difficult to access for most Western researchers. The Chinese and Indian literature, in particular, are difficult to access. For example, we have been hard-pressed to find ancient images of eggplant in India, despite the fact that eggplant is indigenous there. All of our leads so far have come up as dead ends. Thus, the researcher of plant iconography is urged to seek a wide collaboration with scientists in various locations to assist in the quest. Clearly what is needed are databases of plant images of cultivated plants, a cooperative venture between historians, artists, and crop researchers.

REFERENCES

Arber A (1938) Herbals: Their Origin and Evolution. A Chapter in the History of Botany 1470–1670. 2nd ed. Cambridge Univ. Press, Cambridge.

Beck Y (tranl.) (2005) Pedanius Dioscorides of Anazarbus: De Materia Medica, Weidmann, Hildesheim, Germany.

Dane F and J Liu (2007) Diversity and origin of cultivated and citron type watermelon (*Citrullus lanatus*). *Genet Resour Crop Evol* 54:1255-1265.

Darby WJ, P Ghalioungui, L Grivetti (1977) Food: The gift of Osiris, Vol. 2. Academic, London. Figure 17.3, p. 691–695, 717–718, 780–783.

Daunay M-C and J Janick (2007) History and iconography of eggplant. *Chronica Horticulturae* 47(3):16–22.

Daunay M-C and J Janick (2008) Early history and iconography of the Solanaceae: 2. Potato. *SOL Newsletter* 21:5–7.

Daunay M-C, E Jullian, AWhipkey, J Janick (2007a) Eggplant and capsicum peppers: Historical texts and images. Progress in Research in Capsicum & Eggplant. Warsaw Univ. Life Sciences Press, Warsaw, p. 13–21.

Daunay M-C, H Laterrot, J Janick (2007b) Iconography of the Solanaceae from antiquity to the XVIIth century: A rich source of information on genetic diversity and uses. *Acta Horticulturae* 745:59–88.

Daunay M-C, H Laterrot, J Janick (2008) Iconography and history of Solanaceae: Antiquity to the 17th century. *Horticultural Reviews* 34:1–111+31 plates.

Daunay M-C, J Janick, HS Paris (2009) *Tacuinum Sanitatis*: Horticulture and health in the late middle ages. *Chronica Horticulturae* 49(3):22–29.

Der Wiener Dioskurides (1998, 1999) Akademische Druck-u.Verlagsanstalt, 2 Vol., Graz.

Eisendrath ER (1961) Portraits of plants. A limited study of the 'icones'. *Annals of the Missouri Botanical Garden* 48:291–327.

Gerard(e) J (1597) The herbal or general history of plants. Bollifant, London. Republished, 1974, Theatrum Orbis Terrarum, Amsterdam.

Gerard(e) J and T Johnson (1633) The herbal or general history of plants. Islip, Norton and Whitakers, London. Republished, 1974, Dover Publications, Mineola.

Gyulai G, M Humphreys, R Lágler, Z Szabó, Z Tóth, A Bittsánszky, F Gyulai, L Heszky (2006) Seed remains of common millet from the 4th (Mongolia) and 15th (Hungary) centuries: AFLP, SSR and mtDNA sequence recoveries. Seed Science Research 16: 179-191.

Gyulai G, Z Tóth, Z Szabó, F Gyulai, R Lágler, L Kocsis, L Heszky (2009) Domestication Events of Grape (Vitis vinifera) from Antiquity and the Middle Ages in Hungary from growers' viewpoint. Hung Agric Res 2009/3-4: 8-12.

Honig EA (1998) Painting and the Market in Early Modern Antwerp. Yale Univ. Press, New Haven.

Hummer KE and J Janick (2007) Rubus iconography: Antiquity to the Renaissance. Acta Horticulturae 759:89–105.

Janick J (2002) The pear in history, literature, popular culture, and art. Acta Horticulturae 596:41–42.

Janick J (2004a) Caravaggio's fruit. A mirror on Baroque horticulture. Chronica Horticulturae 44(4):9–15.

Janick J (2004b) Erotic use of Lagenaria in Renaissance art. The Cucurbit Network News 11(2):7.

Janick J (2007) Art as a source of information on horticultural technology. Acta Horticulturae 759:69–88.

Janick J and G Caneva (2005) The first images of maize in Europe. Maydica 50:71–80.

Janick J and MC Daunay (2007) History and iconography of the Solanaceae: 1. Mandrake, a plant at the intersection of reason and irrationality. SOL Newsletter 14:3–4.

Janick J and HS Paris (2005) Baby squash in the Italian market, 1580. Cucurbit Network News 12(1):4.

Janick J and HS Paris (2006a) The Cucurbit Images (1515–1518) of the Villa Farnesina, Rome. Annals of Botany 97:165–176.

Janick J and HS Paris (2006b) Old world cucubits in plant iconography of the Renaissance. Proc. Cucurbitaceae 2006, Universal Press, Raleigh, North Carolina. p. 341–348.

Janick J and HS Paris (2006c) Jonah and the "gourd" at Nineveh: Consequences of a classic mistranslation. Proc. Cucurbitaceae. 2006, Universal Press, Raleigh, North Carolina. 349–362.

Janick J, A Whipkey, HS Paris, M-C Daunay, E Julian (2006) Development of an image database of Cucurbitaceae. Proc. Cucurbitaceae 2006, Universal Press, Raleigh, North Carolina. p. 358–362.

Janick J, HS Paris, D Parrish (2007a) The cucurbits of Mediterranean antiquity: Identification of taxa from ancient images and descriptions. Annals of Botany 100:1441–1457.

Janick J, A Whipkey, M-C Daunay, E Jullian, HS Paris (2007b) Development of an image data base of Solanaceae. Acta Horticulturae 745:507–510.

Janick J, M-C, Daunay, HS Paris (2009) Horticulture and health: Ancient Medieval views. International Conference on Horticulture. Nov. 9–12. Bangalore India. p. 23–34.

Janick J, R Kamenetsky, SH Puttaswamy (2010) Horticulture of the Taj Mahal Gardens of the Imagination. Chronica Horticulturae 50(3): in press.

Jashemski WF (1979) The Gardens of Pompeii: Herculaneum and the Villas Destroyed by Versuvius. Caratzas Brothers Publ. New Rochelle, New York.

Keimer L (1924) Die gartenpflanzen im alten Ägypten. Hoffmann & Campe, Hamburg.

Leonard JN (1973) First Farmers. Time Life Books, New York.

Manniche L (1989) An Ancient Egyptian Herbal. Univ. Teas Press. Austin.

Meyer FG, EE Trueblood, JL Heller (1999) The Great Herbal of Leonhart Fuchs. vol. 2, Stanford Univ. Press, Stanford, CA.

Paley SM (1976) King of the world: Ashur-nasir-pal II of Assyria 883–859 B.C. The Brooklyn Museum, New York.

Paris HS and J Janick (2005) Early evidence for the culinary use of squash flowers in Italy. *Chronica Horticulturae* 45(2):20–22.

Paris HS and J Janick (2008a) What the Roman emperor Tiberius grew in his greenhouse. *Cucurbitaceae 2008*. p. 33–41. I.N.R.A., Avignon, France.

Paris HS and J Janick (2008b) Reflections on linguistics as an aid to taxonomical identification of ancient Mediterranean cucurbits: the Piqqus of the Faqqous. *Cucurbitaceae 2008*. p. 43–51. I.N.R.A., Avignon, France.

Paris HS, M-C Daunay, M Pitrat, J Janick (2006a) First known image of *Cucurbita* in Europe, 1503–1508. *Annals of Botany* 98 41–47.

Paris HS, J Janick, M-C Daunay (2006b) First images of *Cucurbita* in Europe. *Proc. Cucurbitaceae 2006*, Universal Press, Raleigh, North Carolina. p. 363–371.

Paris HS, M-C Daunay, J Janick (2009) The Cucurbitaceae and Solanaceae illustrated in medieval manuscripts known as the *Tacuinum Sanitatis*. *Annals of Botany* 103:1187.

Pollack S (1999) Ancient Mesopotamia: The Eden that Never Was. Cambridge Univ. Press, Cambridge, UK.

Renner SS, J Scarborough, H Schaefer, HS Paris, J Janick (2008) Dioscorides's *bruonia melaina* is *Bryonia* alba, not *Tamus communis*, and an illustration labelled bruonia melaina in the Codex Vindobonensis is *Humuls lupulus* not *Bryonia dioica*. *Cucurbitaceae 2008*. p. 273–280. I.N.R.A., Avignon, France.

Singer C, EJ Holmyard, AR Hall (1954) A History of Technology. Vol. 1. Fall of Ancient Empires. Oxford Univ. Press, London.

Tóth Z, G Gyulai, Z Szabó, L Horváth, L Heszky (2007) Watermelon (*Citrullus l. lanatus*) production in Hungary from the Middle Ages (13th cent.). *Hung Agric Res* 2007/4: 14-19.

Tyldesley JA, PG Bahn (1983) Use of plants in the European Paleolithic: A review of the evidence. *Quarternary Science Review*. 2:53–83.

Wellman M (1906–1914) Pedaniio Dioscuridis Anazarbi De material medica libri quinque. Three volumes, reprinted 1958. Weidmann, Berlin.

Zeven AC and WA Brandenburg (1986) Use of paintings from the 16th to the 19th centuries to study the history of domesticated plants. *Econ. Bot.* 40:397–408.

INDEX

E

F